INFERENCE TO THE BEST
EXPLANATION

PHILOSOPHICAL ISSUES IN SCIENCE SERIES
Edited by W. H. Newton-Smith

THE RATIONAL AND THE SOCIAL
James Robert Brown

THE NATURE OF DISEASE
Lawrie Reznek

THE PHILOSOPHICAL DEFENCE OF PSYCHIATRY
Lawrie Reznek

INFERENCE TO THE BEST EXPLANATION
Peter Lipton

SPACE, TIME AND PHILOSOPHY
Christopher Ray

MATHEMATICS AND THE IMAGE OF REASON
Mary Tiles

INFERENCE TO THE BEST EXPLANATION

Peter Lipton

London and New York

First published 1991
by Routledge
11 New Fetter Lane, London EC4P 4EE

Simultaneously published in the USA and Canada
by Routledge
a division of Routledge, Chapman and Hall, Inc.
29 West 35th Street, New York, NY 10001

Typeset in 10/12pt by
LaserScript Limited, Mitcham, Surrey
Printed and bound in Great Britain by
T J Press (Padstow) Ltd, Padstow, Cornwall

British Library Cataloguing in Publication Data
Lipton, Peter
Inference to the best explanation. – (Philosophical
issues in science).
1. Inference
I. Title II. Series
160

Library of Congress Cataloging in Publication Data
Lipton, Peter
Inference to the best explanation / Peter Lipton.
p. cm. — (Philosophical issues in science)
Includes bibliographical references and index.
1. Science—Philosophy. 2. Science—Methodology.
3. Inference. 4. Explanation. I. Title. II. Series.
Q175.L556   1991
501—dc20   90–45766

ISBN 0-415-05886-4

For those who matter most:
my parents, my wife, and my children

CONTENTS

Preface   ix
Introduction   1

1 INDUCTION
Underdetermination   6
Justification   8
Description   13

2 EXPLANATION
Understanding explanation   23
Reason, familiarity, deduction   25

3 THE CAUSAL MODEL
Fact and foil   32
Failed reductions   38
Causal triangulation   41

4 INFERENCE TO THE BEST EXPLANATION
Spelling out the slogan   56
Attractions and repulsions   66

5 CONTRASTIVE INFERENCE
A case study   75
Explanation and deduction   88

6 THE RAVEN PARADOX
Unsuitable contrasts   99
The Method of Agreement   109

vii

CONTENTS

7 LOVELINESS AND INFERENCE
Multiple differences   114
Voltaire's objection   122

8 PREDICTION AND PREJUDICE
The puzzle   133
The fudging explanation   138
Actual and assessed support   150

9 TRUTH AND EXPLANATION
Circularity   158
A bad explanation   168
The scientific evidence   174

Conclusion   185
Bibliography   190
Index   193

PREFACE

It was David Hume's argument against induction that hooked me on philosophy. Surely we have good reason to believe that the sun will rise tomorrow, even though it is possible that it won't; yet Hume provided an apparently unanswerable argument that we have no way to show this. Our inductive practices have been reliable in the past, or we would not be here now to think about them, but an appeal to their past successes to underwrite their future prospects assumes the very practices we are supposed to be justifying. This skeptical tension between my unshakable confidence in the reliability of many of my inferences and the persuasive power of Hume's argument against the possibility of showing any such thing has continued to focus much of my philosophical thinking to this day.

Somewhat later in my education, I was introduced to the problem of description. Even if we cannot see how to justify our inductive practices, surely we can describe them. But I discovered that it is amazingly difficult to give a principled description of the way we weigh evidence. We may be very good at doing it, but we are miserable at describing how it is done, even in broad outline. This book is primarily an investigation of one popular solution to this problem of description, though I also have something to say about its bearing on the problem of justification. It is a solution that looks to explanation as a key to inference, and suggests that we find out what by asking why.

I owe a great debt to the teachers who are my models of how philosophy is to be done and ought to be taught, especially Freddie Ayer, Rom Harré, Peter Harvey, Bill Newton-Smith, and Louis Mink. I have indicated specific obligations to the literature in the body of the text, but there are a number of philosophers

PREFACE

who, through their writing, have pervasively influenced my thinking about inference, explanation, and the relations between them. On the general problem of describing our inductive practices, I owe most to John Stuart Mill, Carl Hempel, and Thomas Kuhn; on the nature of explanation, to Hempel again, Alan Garfinkel, and Michael Friedman; and on Inference to the Best Explanation, to Gilbert Harman.

I am also very pleased to be able to acknowledge the help that many colleagues and friends have given me with the material in this book. In particular, I would like to thank Ken April, Philip Clayton, Matt Ginsberg, Hyman Gross, Jim Hopkins, Colin Howson, Trevor Hussey, Norbet Kremer, Ken Levy, Stephen Marsh, Hugh Mellor, David Papineau, Philip Pettit, Michael Redhead, David Ruben, Mark Sainsbury, Morton Schapiro, Dick Sclove, Dan Shartin, Elliot Sober, Richard Sorabji, Fred Sommers, Ed Stein, Nick Thompson, Laszlo Versenyi, Jonathan Vogel, David Weissbord, Alan White, Jim Woodward, John Worrall, Eddy Zemach, and especially Tim Williamson. All of these people have made this a better book and their philosophical company has been one of the main sources of the pleasure I take in my intellectual life.

I am also grateful to the National Endowment for the Humanities, for a grant under which some of the research for this book was completed, and to Williams College, for a leave during which the final version was written. Several parts of the book are based on previously or soon to be published work, and I thank the editors and publishers in question for permission to use this. Chapter three includes material from 'A Real Contrast' (1987) and from 'Contrastive Explanation' (1991). Chapter eight includes material from 'Prediction and Prejudice' (1990).

Finally, I would like to thank Diana, my wife. Without her, a less readable version of my book would still have been possible, but my life would have been immeasurably poorer.

Peter Lipton
Williamstown, Massachusetts
February, 1990

x

INTRODUCTION

We are forever inferring and explaining, forming new beliefs about the way things are and explaining why things are as we have found them to be. These two activities are central to our cognitive lives, and we usually perform them remarkably well. But it is one thing to be good at doing something, quite another to understand how it is done or why it is done so well. It's easy to ride a bicycle, but very hard to describe how to do it. In the cases of inference and explanation, the contrast between what we can do and what we can describe is stark, for we are remarkably bad at principled description. We seem to have been designed to perform the activities, but not to analyze or defend them. Still, the epistemologist does the best he can with his limited endowment, and tries to describe and justify our inferential and explanatory practices.

This book is an essay on one popular attempt to understand how we go about weighing evidence and making inferences. According to the model of Inference to the Best Explanation, our explanatory practices guide our inferences. Beginning with the evidence available to us, we infer what would, if true, provide the best explanation of that evidence. This cannot be the whole story about inference, but many of our inferences, both in science and in ordinary life, appear to follow this pattern. Faced with tracks in the snow of a certain peculiar shape, I infer that a person on snowshoes has recently passed this way. There are other possibilities, but I make this inference because it provides the best explanation of what I see. Watching me pull my hand away from the stove, you infer that I am in pain, because this is the best explanation of my excited behavior. Having observed the motion of Uranus, the scientist infers that there is another hitherto un-

1

observed planet with a particular mass and orbit, since that is the best explanation of Uranus's path.

Inference to the Best Explanation has become extremely popular, though it also has notable critics. It is widely supposed to provide an accurate description of a central mechanism governing our inferential practices and also a way to show why these practices are reliable. In spite of this, the model has not been much developed. It is more a slogan than an articulated philosophical theory. There has been some discussion of whether this or that inference can be described as to the best explanation, but little investigation into even the most basic structural features of the model. So it is time to try to flesh out the slogan and to give the model the detailed assessment it deserves. That is the purpose of this book.

One reason Inference to the Best Explanation has been so little developed, in spite of its popularity, is clear. The model is an attempt to account for inference in terms of explanation, but our understanding of explanation is so patchy that the model seems to account for the obscure in terms of the equally obscure. It might be correct, yet it is unilluminating. We do not yet have an account that provides the correct demarcation between what explains a phenomenon and what does not; we are even further from an account of what makes one explanation better than another. So the natural idea of articulating Inference to the Best Explanation by plugging in one of the standard theories of explanation yields disappointing results. For example, if we were to insert the familiar deductive-nomological model of explanation, Inference to the Best Explanation would reduce to a variant of the equally familiar hypothetico-deductive model of confirmation. This would not give us a new theory of inference, but only a restatement of an old one that is known to have many weaknesses.

Nevertheless, the situation is far from hopeless. First of all, there are a number of elementary distinctions that can be made which add structure to the account without presupposing any specific and controversial account of explanation. Second, there is some recent and important work on causal explanation and the 'interest relativity' of explanation that can be developed and extended in a way that casts light on the nature of Inference to the Best Explanation and its prospects. Or so I shall try to show.

The book falls into three parts. The first part, chapters one through three, introduces some of the central problems in understanding the nature of inference and of explanation. I distinguish the problems of describing these practices from the problems of justifying them, and consider some of the standard solutions to both. In the third chapter, I focus on contrastive explanations, explanations that answer questions of the form 'Why this *rather than* that?', and attempt an improved account of the way they work.

The second part, chapters four through seven, considers the prospects of Inference to the Best Explanation as a partial solution to the problem of describing our inductive practices. Chapter four develops some of the basic distinctions the model requires, especially the distinction between actual and potential explanation, and between the explanation that is most warranted and the explanation that would, if true, provide the most understanding, the distinction between the likeliest explanation and the loveliest explanation. This chapter also considers some of the *prima facie* strengths and weaknesses of the model. Chapters five and six are an attempt to use the analysis of contrastive explanation from chapter three to defend the model and in particular to show that it marks an improvement on the hypothetico-deductive model of confirmation. In chapter seven, I offer a preliminary extension of the model to exploit other features of explanation and consider some principled objections to the idea that explanation should be a guide to inference.

The third part, chapters eight and nine, examines the prospects of Inference to the Best Explanation as a solution to some of the problems of justifying our inductive practices. Chapter eight considers the use of an explanatory inference to justify the common but controversial view that the successful predictions that a theory makes provide stronger support for it than data that were known before the theory was generated and which the theory was designed to accommodate. The last chapter considers another well-known application of Inference to the Best Explanation, as an argument for scientific realism, where the truth of predictively successful theories is claimed to provide the best explanation of that success. This chapter ends with a brief sketch of some of the prospects for exploiting the structure of scientific inferences in other arguments for realism and against various instrumental interpretations of scientific inference.

The topics of inference and explanation are vast, so there is much that this book assumes and much that it neglects. For example, I help myself to the concept of causation, without offering an analysis of it. My hope is that whatever the correct analysis of that concept turns out to be, it can be plugged into the many claims I make about causal explanation and causal inference without making them false. I also assume throughout that inferred claims, especially inferred theories, are to be construed literally and not, say, by means of some operationalist reduction. For most of the book, I also assume that when a claim is inferred, what is inferred is that the claim is true, or at least approximately true, though this becomes an issue in the final chapter. I do not attempt the difficult task of providing an adequate analysis of the notion of approximate truth or verisimilitude. (For animadversions on this notion and its application, see Laudan (1984) pp. 228–30, and Fine (1984).) Perhaps the most conspicuous absence in the book is any treatment of probabilistic issues, such as Bayesian approaches to inference or statistical models of explanation. I have also neglected the various approaches that workers in Artificial Intelligence have taken to describing inference. These are important matters and ought to be addressed in relation to Inference to the Best Explanation, but I leave them for another time, if not to another person.

I do not count myself a stylish writer, but I can be clear and accessible, at least by the generally low standards of the philosophical literature. So I have tried to write a book that, while not introductory, would be of some use to a dedicated undergraduate unfamiliar with the enormous literature on inference and explanation. As a consequence, some of the material to follow, especially in the first two chapters, will be familiar to *aficionados*. I have also attempted to write so that each chapter stands as much on its own as is compatible with a progressive argument. Consequently, even if you are not particularly interested in Inference to the Best Explanation *per se*, you may find certain chapters worth your time. For example, if you are interested in contrastive explanation, you might just read chapter three, and if you are interested in the prediction/accommodation issue, just chapter eight.

Most philosophers, today and throughout the subject's history, adopt the rhetoric of certainty. They write as if the correctness of their views has been demonstrated beyond reasonable doubt. This sometimes makes for stimulating reading, but it is either disingenuous or naive. In philosophy, if a position is interesting and important, it is almost always also controversial and dubitable. I think Inference to the Best Explanation is both interesting and important, and I have tried not to express more confidence in my claims than the arguments warrant, without at the same time being annoyingly vague or tentative. Since I probably get carried away at points, in spite of my best intentions, let me repeat now that it seems obvious to me that Inference to the Best Explanation cannot be the whole story about inference: at most, it can be an illuminating chapter. And while this book has turned out to take the form of a defense of this model of inference, it should be read rather as a preliminary exploration. Even if it wins no converts, but encourages others to provide more probing criticisms of Inference to the Best Explanation or to generate better alternatives, I will be well satisfied.

1

INDUCTION

UNDERDETERMINATION

Inductive inference is a matter of weighing evidence and judging likelihood, not of proof. How do we go about making these non-demonstrative judgments, and why should we believe they are reliable? Both the question of description and the question of justification arise from underdetermination. To say that an outcome is underdetermined is to say that some information about initial conditions and rules or principles does not guarantee a unique solution. The information that Tom spent five dollars on apples and oranges and that apples are fifty cents a pound and oranges a dollar a pound underdetermines how much fruit Tom bought, given only the rules of deduction. Similarly, those rules and a finite number of points on a curve underdetermine the curve, since there are many curves that would pass through those points.

Underdetermination may also arise in our description of the way a person learns or makes inferences. A description of the evidence, along with a certain set of rules, not necessarily just those of deduction, may underdetermine what is learned or inferred. Insofar as we have described all the evidence and the person is not behaving erratically, this shows that there are hidden rules. We can then study the patterns of learning or inference to try to discover them. Noam Chomsky's argument from 'the poverty of the stimulus' is a good example of how underdetermination can be used to disclose the existence of additional rules (1965, ch. 1, sec. 8, esp. pp. 58–9). Children learn the language of their elders, an ability that enables them to understand an indefinite number of sentences on first acquaintance. The talk

INDUCTION

young children hear, however, along with rules of deduction and any plausible general rules of induction, grossly under-determine the language they learn. What they hear is limited and often includes many ungrammatical sentences, and the little they hear that is well formed is compatible with many possible languages other than the one they learn. Therefore, Chomsky argues, in addition to any general principles of deduction and induction, children must be born with strong linguistic rules or principles that further restrict the class of languages they will learn, so that the actual words they hear are now sufficient to determine a unique language. Moreover, since a child will learn whatever language he is brought up in, these principles cannot be peculiar to a particular human language; instead, they must specify something that is common to all of them. For Chomsky, determining the structure of these universal principles and the way they work is the central task of modern linguistics.

Thomas Kuhn provides another well-known example of using underdetermination as a tool to investigate cognitive principles. He begins from an argument about scientific research strikingly similar to Chomsky's argument about language acquisition (1970; 1977, esp. ch. 12). In most periods in the history of a developed scientific specialty, scientists are in broad agreement about which problems to work on, how to attack them, and what counts as solving them. But the explicit beliefs and rules scientists share, especially their theories, data, general rules of deduction and induction, and any explicit methodological rules, under-determine these shared judgments. Many possible judgments are compatible with these beliefs and rules other than the ones the scientists make. So Kuhn argues that there must be additional field-specific principles that determine the actual judgments. Unlike Chomsky, Kuhn does not argue for principles that are either innate or in the form of rules, narrowly construed. Instead, scientists acquire through their education a stock of exemplars – concrete problem solutions in their specialty – and use them to guide their research. They pick new problems that look similar to an exemplar problem, they try techniques that are similar to those that worked in that exemplar, and they assess their success by reference to the standards of solution that the exemplars illu-strate. Thus the exemplars set up a web of 'perceived similarity relations' that guide future research, and the shared judgments are explained by the shared exemplars. These similarities are not

7

created or governed by rules, but they result in a pattern of research that roughly mimics one that is rule-governed. Just how exemplars do this work, and what happens when they stop working, provide the focus of Kuhn's account of science.

Chomsky and Kuhn are both arguing for unacknowledged principles of induction. In these cases, however, underdetermination is taken to be a symptom of the existence of highly specialized principles, whether of language acquisition or of scientific method in a particular field, since the under-determination is claimed to remain even if we include general principles of induction among our rules. But the same pattern of argument applies to the general case. If an inference is inductive, then by definition it is underdetermined by the evidence and the rules of deduction. Insofar as our inductive practices are methodical, we must use additional rules or principles of inference, and we may study the patterns of our inferences in an attempt to discover them.

JUSTIFICATION

The two central questions about our general principles of induction concern description and justification. What principles do we use, and how can they be shown to be good principles? The question of description seems at first to take priority. How can we even attempt to justify our principles until we know what they are? Historically, however, the justification question came first. One reason for this is that this question gets its grip from skeptical arguments that seem to apply to any principles that could account for the way we fill the gap between evidence and inference.

The problem of justification is to show that our inferential methods are good methods. The natural way to understand this is in terms of truth. We want our methods of inference to be 'truth-tropic', to take us towards the truth. For deduction, a good argument is one that is valid, a perfect truth conduit, where it is impossible for there to be true premises but a false conclusion. The problem of justification here would be to show that arguments we judge valid are in fact so. For induction, perfect reliability is out of the question. By definition, even a good inductive argument is one where it is possible for there to be true premises but a false conclusion. Moreover, it is clear that the reasonable inductive inferences we make are not 100 per cent reliable even in this world, since they sometimes sadly take us from truth to falsehood. Nevertheless, it remains natural to construe the task of justification as that of showing truth-tropism. We would like to show that those inductive inferences we judge worth making are ones that usually take us from true premises to true conclusions.

A skeptical argument that makes the problem of justification pressing has two components, underdetermination and circularity. The first is an argument that the inferences in question are underdetermined, given only our premises and the rules of deduction; that the premises and those rules are compatible not just with the inferences we make, but also with other, incompatible inferences. This shows that the inferences we make really are inductive and, by showing that there are possible worlds where the principles we use take us from true premises to false conclusions, it also shows that there are worlds where our principles would fail us. Revealing this underdetermination, however, does not yet generate a skeptical argument, since we might have good reason to believe that the actual world is one where our principles are reliable. So the skeptical argument requires a second component, an argument for circularity, which attempts to show that we cannot rule out the possibility of unreliability that underdetermination raises without employing the very principles that are under investigation, and so begging the question.

Although it is not traditionally seen as raising the problem of induction, Descartes' 'First Meditation' is a classic illustration of this technique. Descartes' goal is to cast doubt on the 'testimony of the senses', which leads us to infer that there is, say, a mountain in the distance because that is what it looks like. He begins by arguing that we ought not to trust the senses completely, since we know that they do sometimes mislead us, 'when it is a question of very small and distant things'. This argument relies on the underdetermination component, since it capitalizes on the fact that the way things appear does not entail the way they are, but it does not yet have the circle component. We can corroborate our inferences about small and distant things without circularity by taking a closer look (cf. Williams, 1978, pp. 51–2). But Descartes immediately moves on from the small and the distant to the large and near. No matter how clearly we seem to see something, it

9

may only be a dream, or a misleading experience induced by an evil demon. These arguments describe possible situations where even the most obvious testimony of the senses is misleading. Moreover, unlike the worry about small and distant things, these arguments also have a circle component. There is apparently no way to test whether a demon is misleading us with a particular experience, since any test would itself rely on experiences that he might have induced. The senses may be liars and give us false testimony, and we should not find any comfort if they also report that they are telling us the truth.

The demon argument begins with the underdetermination of observational belief by observational experience, construes the missing principle of inference on the model of inference from testimony, and then suggests that the reliability of this principle could only be shown by assuming it. Perhaps one of the reasons Descartes' arguments are not traditionally seen as raising the problem of justifying induction is that his response to his own skepticism is to reject the underdetermination upon which it rests. Descartes argues that, since inferences from the senses must be inductive and so raise a skeptical problem, our knowledge must instead have a different sort of foundation for which the problem of underdetermination does not arise. The *cogito* and the principles of clearness and distinctness that it exemplifies are supposed to provide the non-inductive alternative. In other words, the moral he draws from underdetermination and circularity is not that our principles of induction require some different sort of defense or must be accepted without justification, but that we must use different premises and principles, for which the skeptical problem does not arise. For a skeptical argument about induction that does not lead to the rejection of induction, we must turn to its traditional home, in the arguments of David Hume.

Hume also begins with underdetermination, in this case that our observations do not entail our predictions (1777, sec. IV). He then suggests that the governing principle of all our inductive inferences is that nature is uniform, that the unobserved (but observable) world is much like what we have observed. The question of justification is then the question of showing that nature is indeed uniform. This cannot be deduced from what we have observed, since the claim of uniformity itself incorporates a massive prediction. But the only other way to argue for uniformity is to use an inductive argument, which would rely on the principle of uniformity, leaving the question begged. According to Hume, we are addicted to the practice of induction, but it is a practice that cannot be justified.

To illustrate the problem, suppose our fundamental principle of inductive inference is 'More of the Same'. We believe that strong inductive arguments are those whose conclusions predict the continuation of a pattern described in the premises. Applying this principle of conservative induction, we would infer that the sun will rise tomorrow, since it has always risen in the past; and we would judge worthless the argument that the sun will not rise tomorrow since it has always risen in the past. It is, however, easy to come up with a factitious principle to underwrite the latter argument. According to the principle of revolutionary induction, 'It's Time for a Change', and this sanctions the dark inference. Hume's argument is that we have no way to show that conservative induction, the principle he claims we actually use for our inferences, will do any better than intuitively wild principles like the principle of revolutionary induction. Of course conservative induction has had the more impressive track record. Most of the inferences from true premises that it has sanctioned have also had true conclusions. Revolutionary induction, by contrast, has been conspicuous in failure, or would have been, had anyone relied on it. The question of justification, however, does not ask which method of inference has been successful; it asks which one will be successful.

Still, the track record of conservative induction appears to be a reason to trust it. That record is imperfect (we are not aspiring to deduction), but very impressive, particularly as compared with revolutionary induction and its ilk. In short, induction will work because it has worked. This seems the only justification our inductive ways could ever have or require. Hume's disturbing observation was that this justification appears circular, no better than trying to convince someone that you are honest by saying that you are. Much as Descartes argued that we should not be moved if the senses give testimony on their own behalf, so Hume argued that we cannot appeal to the history of induction to certify induction. The trouble is that the argument that conservative inductions will work because they have worked is itself an induction. The past success is not supposed to prove future success, only make it very likely. But then we must decide which

11

standards to use to evaluate this argument. It has the form 'More of the Same', so conservatives will give it high marks, but since its conclusion is just to underwrite conservatism, this begs the question. If we apply the revolutionary principle, it counts as a very weak argument. Worse still, by revolutionary standards, conservative induction is likely to fail precisely because it has succeeded in the past, and the past failures of revolutionary induction augur well for its future success (cf. Skyrms, 1986, ch. 2). The justification of revolutionary induction seems no worse than the justification of conservative induction, which is to say that the justification of conservative induction looks very bad indeed.

The problem of justifying induction does not show that there are other inductive standards better than our own. Instead it argues for a deep symmetry: many sets of standards, most of them wildly different from our own and incompatible with each other, are yet completely on a par from a justificatory point of view. This is why the problem of justification can be posed before we have solved the problem of description. Whatever inductive principles we use, the fact that they are inductive seems enough for the skeptic to show that they defy justification. We fill the gap of underdetermination between observation and prediction in one way, but it could be filled in many other ways that would have led to entirely different predictions. We have no way of showing that our way is any better than any of these others. It is not merely that the revolutionaries will not be convinced by the justificatory arguments of the conservatives: the conservatives should not accept their own defense either, since among their standards is one which says that a circular argument is a bad argument, even if it is in one's own aid. Even if I am honest, I ought to admit that the fact that I say so ought not carry any weight. We have a psychological compulsion to favor our own inductive principles but, if Hume is right, we should see that we cannot even provide a cogent rationalization of our behavior.

It seems to me that we do not yet have a satisfying solution to Hume's challenge and that the prospects for one are bleak, but there are other problems of justification that are more tractable. The peculiar difficulty of meeting Hume's skeptical argument against induction is that he casts doubt on our inductive principles as a whole, and so any recourse to induction to justify induction seems hopeless. But one can also ask for the justification of particular inductive principles and, as Descartes' example of small and distant things suggests, this leaves open the possibility of appeal to other principles without begging the question. For example, among our principles of inference is one that makes us more likely to infer a theory if it is supported by a variety of evidence than if it is supported by a similar amount of homogeneous data. This is the sort of principle that might be justified in terms of a more basic inductive principle, say that we have better reason to infer a theory when all the reasonable competitors have been refuted, or that a theory is only worth inferring when each of its major components has been separately tested. Another, more controversial, example of a special principle that might be justified without circularity is that, all else being equal, a theory deserves more credit from its successful predictions than it does from data that the theory was constructed to fit. This appears to be an inductive preference most of us have, but the case is controversial because it is not at all obvious that it is rational. On the one hand, many people feel that only a prediction can be a real test, since a theory cannot possibly be refuted by data it is built to accommodate; on the other, that logical relations between theory and data upon which inductive support exclusively depends cannot be affected by the merely historical fact that the data were available before or only after the theory was proposed. In any event, this is an issue of inductive principle that is susceptible to non-circular evaluation, as we will see in chapter eight. Finally, though a really satisfying solution to Hume's problem would have to be an argument for the reliability of our principles that had force against the inductive skeptic, there may be arguments for reliability that do not meet this condition yet still have probative value for those of us who already accept some forms of induction. We will consider some candidates in chapter nine.

DESCRIPTION

We can now see why the problem of justification, the problem of showing that our inductive principles are reliable, did not have to wait for a detailed description of those principles. The problem of justifying our principles gets its bite from skeptical arguments, and these appear to depend only on the fact that these principles are principles of induction, not on the particular form they take.

The crucial argument is that the only way to justify our principles would be to reason with the very same principles, which is illegitimate; an argument that seems to work whatever the details of our inferences. The irrelevance of the details comes out in the symmetry of Hume's argument: just as the future success of conservative induction gains no plausibility from its past success, so the future success of revolutionary induction gains nothing from its past failures. As the practice varies, so does the justificatory argument, preserving the pernicious circularity. Thus the question of justification has had a life of its own: it has not waited for a detailed description of the practice whose warrant it calls into doubt.

By the same token, the question of description has fortunately not waited for an answer to the skeptical arguments. Even if our inferences were unjustifiable, one still might be interested in saying how they work. The problem of description is not to show that our inferential practices are reliable; it is simply to describe them as they stand. One might have thought that this would be a relatively trivial problem. First of all, there are no powerful reasons for thinking that the problem is impossible, as there are for the problem of justification. There is no great skeptical argument against the possibility of description. It is true that any account of our principles will require inductive support, since we must see whether it jibes with our observed inductive practice. This, however, raises no general problem of circularity now that justification is not the issue. Using induction to investigate induction is no more a problem here than using observation to study the structure and function of the eye. Second, it is not just that a solution to the problem of describing our inductive principles should be possible, but that it should be easy. After all, they are our principles, and we use them constantly. It thus comes as something of a shock to discover how extraordinarily difficult the problem of description has turned out to be. It is not merely that ordinary reasoners are unable to describe what they are doing: at least one hundred and fifty years of focused effort by epistemologists and philosophers of science has yielded little better. Again, it is not merely that we have yet to capture all the details, but that the most popular accounts of the gross structure of induction are wildly at variance with our actual practice.

Why is description so hard? One reason is a quite general gap between what we can do and what we can describe. You may know how to do something without knowing how you do it; indeed, this is the usual situation. It is one thing to know how to tie one's shoes or to ride a bike; it is quite another thing to be able to give a principled description of what it is that one knows. Chomsky's work on principles of language acquisition and Kuhn's work on scientific method are good cognitive examples. Their investigations would not be so important and controversial if the ordinary speaker of a language knew how she distinguished grammatical from ungrammatical sentences or the normal scientist knew how he made his methodological judgments. The speaker and the scientist employ various principles, but they are not conscious of them. The situation is similar in the case of inductive inference generally. Although we may partially articulate some of our inferences if, for example, we are called upon to defend them, we are not conscious of the fundamental principles of inductive inference we constantly use.

Since our principles of induction are neither available to introspection, nor observable in any other way, the evidence for their structure must be highly indirect. The project of description is one of black box inference, where we try to reconstruct the underlying mechanism on the basis of the superficial patterns of evidence and inference we observe in ourselves. This is no trivial problem. Part of the difficulty is simply the fact of underdetermination. As the examples of Chomsky and Kuhn show, underdetermination can be a symptom of missing principles and a clue to their nature, but it is one that does not itself determine a unique answer. In other words, where the evidence and the rules of deduction underdetermine inference, that information also underdetermines the missing principles. There will always be many different possible mechanisms that would produce the same patterns, so how can one decide which one is actually operating? In practice, however, we usually have the opposite problem: we can not come up with even one description that would yield the patterns we observe. The situation is the same in scientific theorizing generally. There is always more than one account of the unobserved and often unobservable world that would account for what we observe, but scientists' actual difficulty is often to come up with even one theory that fits the observed facts. On reflection, then, it should not surprise us that the problem of description has turned out to be so difficult. Why should we suppose that the project of describing our inductive principles is going to be easier than it would be, say, to give a detailed account of the working of a computer on the basis of the correlations between keys pressed and images on the screen?

Now that we are prepared for the worst, we may turn to some of the popular attempts at description. In my discussion of the problem of justification, I suggested that, following Hume's idea of induction as habit formation, we describe our pattern of inference as 'More of the Same'. This is pleasingly simple, but the conservative principle is at best a caricature of our actual practice. We sometimes do not infer that things will remain the same and we sometimes infer that things are going to change. When my mechanic tells me that my brakes are about to fail, I do not suppose that he is therefore a revolutionary inductivist. Again, we often make inductive inferences from something we observe to something invisible, such as from people's behavior to their beliefs or from the scientific evidence to unobservable entities and processes, and this does not fit into the conservative mold. 'More of the Same' might enable me to predict what you will do on the basis of what you have done (if you are a creature of habit), but it will not tell me what you are or will be thinking.

Faced with the difficulty of providing a general description, a reasonable strategy is to begin by trying to describe one part of our inductive practice. This is a risky procedure, since the part one picks may not really be describable in isolation, but there are sometimes reasons to believe that a particular part is independent enough to permit a useful separation. Chomsky must believe this about our principles of linguistic inference. Similarly, one might plausibly hold that, while simple habit formation cannot be the whole of our inductive practice, it is a core mechanism that can be treated in isolation. Thus one might try to salvage the intuition behind the conservative principle by giving a more precise account of the cases where we are willing to project a pattern into the future, leaving to one side the apparently more difficult problems of accounting for predictions of change and inferences to the unobservable. What we may call the instantial model of inductive confirmation may be seen in this spirit. According to it, a hypothesis of the form 'All A's are B' is supported by its positive instances, by observed A's that are also B (cf. Hempel, 1965, ch. 1). This is not, strictly speaking, an account of inductive *inference*, since it does not say either how we come up with the hypothesis in the first place or how many supporting instances are required

16

before we actually infer it, but this switching of the problem from inference to support may also be taken as a strategic simplification. In any event, the underlying idea is that, if enough positive instances and no refuting instances (A's that are not B) are observed, we will infer the hypothesis, from which we may then deduce the prediction that the next A we observe will be B.

This model could only be a very partial description of our inductive principles but, within its restricted range, it strikes many people initially as a truism, and one that captures Hume's point about our propensity to extend observed patterns. Observed positive instances are not necessary for inductive support, as inferences to the unobserved and to change show, but they might seem at least sufficient. The instantial model, however, has been shown to be wildly over-permissive. Some hypotheses are supported by their positive instances, but many are not. Observing only black ravens may lead one to believe that all ravens are black, but observing only bearded philosophers would probably not lead one to infer that all philosophers are bearded. Nelson Goodman has generalized this problem, by showing how the instantial model sanctions any prediction at all if there is no restriction on the hypotheses to which it can be applied (Goodman, 1983, ch. 3). His technique is to construct hypotheses with factitious predicates. Black ravens provide no reason to believe that the next swan we see will be white, but they do provide positive instances of the artificial hypothesis that 'All raveswans are blight', where something is a raveswan just in case it is either observed before today and a raven, or not so observed and a swan, and where something is blight just in case it is either observed before today and black, or not so observed and white. But the hypothesis that all raveswans are blight entails that the next observed raveswan will be blight which, given the definitions, is just to say that the next swan will be white.

The other famous difficulty facing the instantial model arises for hypotheses that do seem to be supported by their instances. Black ravens support the hypothesis that all ravens are black. This hypothesis is logically equivalent to the contrapositive hypothesis that all non-black things are non-ravens: there is no possible situation where one hypothesis would be true but the other false. According to the instantial model, the contrapositive hypothesis is supported by non-black non-ravens, such as green leaves. The rub comes with the observation that whatever supports a hypothesis also supports anything logically equivalent to it. This is very plausible, since support provides a reason to believe true, and we know that if a hypothesis is true, then so must be anything logically equivalent to it. But then the instantial model once again makes inductive support far too easy, counting green leaves as evidence that all ravens are black (Hempel, 1965, ch. 1). We will discuss this paradox of the ravens in chapter six.

Another account of inductive support is the hypothetico-deductive model (Hempel, 1966, chs 2, 3). On this view, a hypothesis or theory is supported when it, along with various other statements, deductively entails a datum. Thus a theory is supported by its successful predictions. This account has a number of attractions. First, although it leaves to one side the important question of the source of hypotheses, it has much wider scope than the instantial model, since it allows for the support of hypotheses that appeal to unobservable entities and processes. The big bang theory of the origin of the universe obviously cannot be directly supported but, along with other statements, it entails that we ought to find ourselves today traveling through a uniform background radiation, like the ripples left by a rock falling into a pond. The fact that we do now observe this radiation (or effects of it) provides some reason to believe the big bang theory. Thus, even if a hypothesis cannot be supported by its instances, because its instances are not observable, it can be supported by its observable logical consequences. Second, the model enables us to co-opt our accounts of deduction for an account of induction, an attractive possibility since our understanding of deductive principles is so much better than our understanding of inductive principles. Lastly, the hypothetico-deductive model seems genuinely to reflect scientific practice, which is perhaps why it has become the scientists' philosophy of science.

In spite of all its attractions, our criticism of the hypothetico-deductive model here can be brief, since it inherits all the over-permissiveness of the instantial model. Any case of support by positive instances will also be a case of support by consequences. The hypothesis that all A's are B, along with the premise that an individual is A, entails that it will also be B, so the thing observed to be B supports the hypothesis, according to the hypothetico-deductive model. That is, any case of instantial support is also a case of hypothetico-deductive support, so the model has to face the problem of insupportable hypotheses and the raven paradox. Moreover, the hypothetico-deductive model is similarly over-permissive in the case of vertical inferences to hypotheses about unobservables, a problem that the instantial model avoided by ignoring such inferences altogether. The difficulty is structurally similar to Goodman's problem of factitious predicates. Consider the conjunction of the hypotheses that all ravens are black and that all swans are white. This conjunction, along with premises concerning the identity of various ravens, entails that they will be black. According to the model, the conjunction is supported by black ravens, and it entails its own conjunct about swans. The model thus appears to sanction the inference from black ravens to white swans (cf. Goodman, 1983, pp. 67–8). Similarly, the hypothesis that all swans are white taken alone entails the inclusive disjunction that either all swans are white or there is a black raven, a disjunction we could establish by seeing a black raven, thus giving illicit hypothetico-deductive support to the swan hypothesis. These maneuvers are obviously artificial, but nobody has managed to show how the model can be modified to avoid them without also eliminating most genuine cases of inductive support (cf. Glymour, 1980, ch. 2). Finally, in addition to being too permissive, finding support where none exists, the model is also too strict, since data may support a hypothesis which does not, along with reasonable auxiliary premises, entail them. We will investigate this problem in chapter five and return to the problem of over-permissiveness in chapter six.

We have now canvassed three attempts to tackle the descriptive problem: 'More of the Same', the instantial model, and the hypothetico-deductive model. The first two could, at best, account for a special class of particularly simple inferences, and all three are massively over-permissive, finding inductive support where there is none to be had. They do not give enough structure to the black box of our inductive principles to determine the inferences we actually make. This is not to say that these accounts describe mechanisms that would yield too many inferences: they would probably yield too few. A 'hypothetico-deductive box', for example, would probably have little or no inferential output, given the plausible additional principle that we will not make inferences we know to be contradictory. For every hypothesis that we would be inclined to infer on the basis of the deductive support it enjoys, there will be an incompatible hypothesis that is similarly supported, and the result is no inference at all, so long as both hypotheses are considered and their incompatibility recognized.

A fourth account of induction, the last I will consider in this section, focuses on causal inference. It is a striking fact about our inductive practice, both lay and scientific, that so many of our inferences depend on inferring from effects to their probable causes. This is something that Hume himself emphasized (Hume, 1777, sec. IV). The cases of direct causal inference are legion, such as the doctor's inference from symptom to disease, the detective's inference from evidence to perpetrator, the mechanic's inference from the engine noises to what is broken, and many scientific inferences from data to theoretical explanation. Moreover, it is striking that we often make a causal inference even when our main interest is in prediction. Indeed, the detour through causal theory on the route from data to prediction seems to be at the center of many of the dramatic successes of scientific prediction. All this suggests that we might do well to consider an account of the way causal inference works as a central component of a description of our inductive practice.

The best known account of causal inference is John Stuart Mill's discussion of the 'methods of experimental inquiry' (Mill, 1904, book III, ch. VIII). The two central methods are the Method of Agreement and especially the Method of Difference. According to the Method of Agreement, in idealized form, when we find that there is only one antecedent that is shared by all the observed instances of an effect, we infer that it is a cause (book III, ch. VIII, sec. 1; references to Mill hereafter as, e.g., 'III.VIII.1'). Thus we come to believe that hangovers are caused by heavy drinking. According to the Method of Difference, when we find that there is only one prior difference between a situation where the effect occurs and an otherwise similar situation where it does not, we infer that the antecedent that is only present in the case of the effect is a cause (III.VIII.2). If we add sodium to a blue flame, and the flame turns yellow, we infer that the presence of sodium is a cause of the new color, since that is the only difference between the flame before and after the sodium was added. If we once successfully follow a recipe for baking bread, but fail another time because we have left out the yeast and the bread does not rise, we would infer that the yeast is a cause of the rising in the first case. Both methods work by a combination of retention and variation. When we apply the Method of Agreement, we hold the effect constant and try to vary the background as much as we can, and see what stays the same; when we apply the Method of Difference, we vary the effect, and try to hold as much of the background constant as we can, and see what changes.

Mill's methods have a number of attractive features. Many of our inferences are causal inferences, and Mill's methods give a natural account of these. In science, for example, the controlled experiment is a particularly common and self-conscious application of the Method of Difference. The Millian structure of causal inference is often particularly clear in cases of inferential dispute. When you dispute my claim that C is the cause of E, you will often make your case by pointing out that the conditions for Mill's methods are not met; that is, by pointing out C is not the only antecedent common to all cases of E, or that the presence of C is not the only salient difference between a case where E occurs and a similar case where it does not. Mill's methods may also avoid some of the over-permissiveness of other accounts, because of the strong constraints that the requirements of varied or shared backgrounds place on their application. These requirements suggest how our background beliefs influence our inferences, something a good account of inference must do. The methods also help to bring out the roles in inference of competing hypotheses and negative evidence, as we will see in chapter five, and the role of background knowledge, as we will see in chapter seven.

Of course, Mill's methods have their share of liabilities, of which I will mention just two. First, they do not themselves apply to unobservable causes or to any causal inferences where the cause's existence, and not just its causal status, is inferred. Second, if the methods are to apply at all, the requirement that there be only a single agreement or difference in antecedents must be seen as an idealization, since this condition is never met in real life. We need principles for selecting from among multiple agreements or similarities those that are likely to be causes, but these are principles Mill does not himself supply. As we will see in chapters five through seven, however, Mill's method can be modified and expanded in a way that may avoid these and other liabilities it faces in its simple form.

This chapter has set part of the stage for an investigation of our inductive practices. I have suggested that many of the problems those practices raise can be set out in a natural way in terms of the underdetermination that is characteristic of inductive inference. The underdetermination of our inferences by our evidence provides the skeptic with his lever, and so poses the problem of justification. It also elucidates the structure of the descriptive problem, and the black box inferences it will take to solve it. I have canvassed several solutions to the problem of description, partly to give a sense of some of our options and partly to suggest just how difficult the problem is. But at least one solution to the descriptive problem was conspicuous by its absence, the solution that gives this book its title and which will be at the center of attention from chapter four onwards. According to Inference to the Best Explanation, we infer what would, if true, be the best explanation of our evidence. On this view, explanatory considerations are our guide to inference. So to develop and assess this view, we need first to look at another sector of our cognitive economy, our explanatory practices. This is the subject of the next two chapters.

2

EXPLANATION

UNDERSTANDING EXPLANATION

Once we have made an inference, what do we do with it? Our inferred beliefs are guides to action that help us to get what we want and avoid trouble. Less practically, we also sometimes infer simply because we want to learn more about the way the world is. Often, however, we are not satisfied to discover that something is the case: we want to know *why*. Thus our inferences may be used to provide explanations, and they may themselves be explained. The central question about our explanatory practices can be construed in several ways. We may ask what principles we use to distinguish between a good explanation, a bad explanation, and no explanation at all. Or we may ask what relation is required between two things to count one to be an explanation of the other. We can also formulate the question in terms of the relationship between knowledge and understanding. Typically, someone who asks why something is the case already knows that it is the case. The person who asks why the sky is blue knows that it is blue, but does not yet understand why. The question about explanation can then be put this way: What has to be added to knowledge to get understanding?

As in the case of inference, explanation raises problems both of justification and of description. The problem of justification can be understood in various ways. It may be seen as the problem of showing whether things we take to be good explanations really are, whether they really provide understanding. The issue here, to distinguish it from the case of inference, is not whether there is any reason to believe that our putative explanations are themselves true, but whether, granting that they are true, they really

23

explain. There is no argument against the possibility of explanation on a par with Hume's argument against induction. The closest thing is regress of why. The why-regress is a feature of the logic of explanation that many of us discovered as children, to our parents' cost. I vividly recall the moment it dawned on me that, whatever my mother's answer to my latest why-question, I could simply retort by asking 'Why?' of the answer itself, until my mother ran out of answers or patience. But if only something that is itself understood can be used to explain, and understanding only comes through being explained by something else, then the infinite chain of why's makes explanation impossible. Sooner or later, we get back to something unexplained, which ruins all the attempts to explain that are built upon it (cf. Friedman, 1974, pp. 18–19).

This skeptical argument is not very troubling. One way to stop the regress is to argue for phenomena that are self-explanatory or that can be understood without explanation. But while there may be such phenomena, this reply concedes too much to the skeptical argument. A better reply is that explanations need not themselves be understood. A drought may explain a poor crop, even if we don't understand why there was a drought; I understand why you didn't come to the party if you explain that you had a bad headache, even if I have no idea why you had a headache; the big bang explains the background radiation, even if the big bang is itself inexplicable, and so on. Understanding is not like a substance that the explanation has to possess in order to pass it on to the phenomenon to be explained. Rather than show that explanation is impossible, the regress argument brings out the important facts that explanations can be chained and that what explains need not itself be understood, and so provides useful constraints on a proper account.

The reason there is no skeptical argument against explanation on a par with Hume's argument against induction is fairly clear. Hume's argument, like all the great skeptical arguments, depends on our ability to see how our methods of acquiring beliefs could lead us into massive error. There are possible worlds where our methods persistently mislead us, and Hume exploits these possibilities by arguing that we have no non-circular way of showing that the actual world is not one of these misleading worlds. In the case of explanation, by contrast, the skeptic does not have this handle, at least when the issue is not whether the

EXPLANATION

explanation is true, but whether the truth really explains. We do not know how to make the contrast between understanding and merely seeming to understand in a way that makes sense of the possibility that most of the things that meet all our standards for explanation might nonetheless not really explain. To put the matter another way, we do not see a gap between meeting our standards for the explanation and actually understanding in the way we easily see a gap between meeting our inductive standards and making an inference that is actually correct.

It is not clear whether this is good or bad news for explanation. On the one hand, in the absence of a powerful skeptical argument, we feel less pressure to justify our practice. On the other, the absence seems to show that our grasp on explanation is even worse than our grasp on inference. We know that inferences are supposed to take us to truths and, as Hume's argument illustrates, we at least have some appreciation of the nature of these ends independently of the means we use to try to reach them. The situation is quite different for explanation. We may say that understanding is the goal of explanation, but we do not have a clear conception of understanding apart from whatever our explanations provide. If this is right, the absence of powerful skeptical arguments against explanation does not show that we are in better shape here than we are in the case of inference. Perhaps things are even worse for explanation: here we may not even know what we are *trying* to do. Once we know that something is the case, what is the point of asking why?

REASON, FAMILIARITY, DEDUCTION

Explanation also raises the problem of description. Whatever the point of explanation or the true nature of understanding, we have a practice of giving and judging explanations, and the problem of description is to give an account of how we do this. As with the problem of justification, the central issue here is not how we judge whether what we claim to be an explanation is true, but whether, granting that it is true, it really does explain what it purports to explain. Like our inductive practices, our explanatory practices display the gap between doing and describing. We discriminate between things we understand and things we do not, and between good explanations and bad explanations, but we are strikingly bad at giving any sort of principled account of

25

how we do this. As before, the best way to make this claim convincing is to canvass some of the objections to various popular accounts of explanation. In this chapter, I will consider three accounts: the reason model, the familiarity model, and the deductive-nomological model.

According to the reason model, to explain a phenomenon is to give a reason to believe that the phenomenon occurs (cf. Hempel 1965, pp. 337, 364–76). On this view, an engineer's explanation of the collapse of a bridge succeeds by appealing to theories of loading, stress, and fatigue which, along with various particular facts, show that the collapse was likely. There is a germ of truth in this view, since explanations do quite often make the phenomenon likely and give us a reason to believe it occurs. A particularly satisfying type of explanation takes a phenomenon that looks accidental and shows how, given the conditions, it was really inevitable, and these deterministic explanations do seem to provide strong reasons for belief. Moreover, the reason model suggests a natural connection between the explanatory and predictive uses of scientific theories since, in both cases, the theory works by providing grounds for belief.

On balance, however, the reason model is extremely implausible. It does not account for the central difference between knowing that a phenomenon occurs and understanding why it occurs. The model claims that understanding why the phenomenon occurs is having a reason to believe that it occurs, but we already have this when we know that it occurs. We already have a reason to believe the bridge collapsed when we ask why it did, so we can not simply be asking for a reason when we ask for an explanation. Explanations may provide reasons for belief, but that is not enough. It is also often too much: many explanations do not provide any actual reason to believe that the phenomenon occurs. Suppose you ask me why there are certain peculiar tracks in the snow in front of my house. Looking at the tracks, I explain to you that a person on snowshoes recently passed this way. This is a perfectly good explanation, even if my only reason for believing it is that I see the very tracks I am explaining. This 'self-evidencing explanation' (Hempel 1965, pp. 370–4) has a distinctive circularity: the passing person on snowshoes explains the tracks and the tracks provide the evidence for the passing. What is significant is that this circularity is *virtuous*: it ruins neither the explanation nor the justification. It does, however, show that the justification view of explanation is false, since to take the explanation to provide a reason to believe the phenomenon after the phenomenon has been used as a reason to believe the explanation would be vicious. In other words, if the reason model were correct, self-evidencing explanations would be illicit, but self-evidencing explanations may be perfectly acceptable and are ubiquitous, as we will see in chapter four, so the reason model is wrong. Providing reasons for belief is neither necessary nor sufficient for explanation.

Another answer to the descriptive question is the familiarity model. On this view, unfamiliar phenomena call out for explanation, and good explanations somehow make them familiar (cf. Hempel, 1965, pp. 430–3; Friedman, 1974, pp. 9–11). On a common version of this view, there are certain familiar phenomena and processes that we do not regard as in need of explanation, and a good explanation of a phenomenon not in this class consists in showing it to be the outcome of a process that is analogous to the processes that yield the familiar phenomena. The kinetic theory of gases explains various phenomena of heat by showing that gases behave like collections of tiny billiard balls; Darwin's theory of natural selection explains the traits of plants and animals by describing a mechanism similar to the mechanism of artificial selection employed by animal breeders; and a theory of electronics explains by showing that the flow of current through a wire is like the flow of water through pipes. But this is not a particularly attractive version of the familiarity view, in part because it does not help us to understand why certain phenomena are familiar in the first place, and because not all good explanations rely on analogies.

A more promising version of the familiarity view begins with the idea that a phenomenon is unfamiliar when, although we may know that it occurs, it remains surprising because it is in tension with other beliefs we hold. A good explanation shows how the phenomenon arises in a way that eliminates the tension and so the surprise (cf. Hempel, 1965, pp. 428–30). We may know that bats navigate with great accuracy in complete darkness, yet find this very surprising, since it seems in tension with our belief that vision is impossible in the dark. Finding out about echolocation shows that there is no real tension, and we are no longer surprised. The magician tells me the number I was thinking of, to my great surprise; a good explanation of the trick

27

ruins it by making it unsurprising. This version of the familiarity view has the virtue of capturing the fact that it is very often surprise that prompts the search for explanations. It also calls our attention to the process of 'defamiliarization', which is often the precursor to asking why various common phenomena occur. In one sense, the fact that the sky is blue is a paradigm of familiarity, but we become interested in explaining this when we stop to consider how odd it is that the sky should have any color at all. Again, the fact that the same side of the moon always faces us at first seems not to call out for any interesting explanation, since it just seems to show that the moon is not spinning on its own axis. It is only when we realize that the moon must actually spin in order to keep the same side towards us, and moreover with a period that is exactly the same as the period of its orbit around the earth, that this phenomenon cries out for explanation. The transformation of an apparently familiar phenomenon into an extraordinary 'coincidence' prompts the search for an adequate explanation. The surprise version of the familiarity model also suggests that a good explanation of a phenomenon will some- times show that beliefs that made the phenomenon surprising are themselves in error, and this is an important part of our explanatory practice. If I am surprised to see a friend at the supermarket because I expected him to be away on vacation, he will not satisfy my curiosity simply by telling me that he needed some milk, but must also say something about why my belief about his travel plans was mistaken.

There are three objections that are commonly made to a familiarity model of explanation. One is that familiarity is too subjective a notion to yield a suitably objective account of explanation. The surprise version of the familiarity model does make explanation audience relative, since what counts as a good explanation will depend on prior expectations, which vary from person to person. But it is not clear that this is a weakness of the model. Nobody would argue that an account of inference is unacceptable because it makes warranted inductions vary with prior belief. Similarly, an account that makes explanation interest relative does not thereby make explanation perniciously subjective. (We will consider the interest relativity of explanation in the next chapter.) In particular, the familiarity model does not collapse the distinction between understanding a phenomenon and mistakenly thinking one does. The explanation must itself be true, and it must say something about how the phenomenon came about, but just what it says may legitimately depend on the audience's interests and expectations.

Another objection to the familiarity model is that good explanations often themselves appeal to unfamiliar events and processes. This is particularly common in the advanced sciences. Here again, however, it is unclear whether this is really a difficulty for a reasonable version of the model. Scientific explanations often appeal to exotic processes, but it is not clear that such an explanation is very good if it conflicts with other beliefs. The third and most telling objection is that we often explain familiar phenomena. The process of defamiliarization goes some way towards meeting this, but it is not enough to turn the objection completely. The rattle in my car is painfully familiar, and consistent with everything else I believe, but while I am sure there is a good explanation for it, I don't have any idea what it is. Nor do you have to convince me that, in fact, it is somehow surprising that there should be a rattle to get me interested in explaining it. Surprise is often a precursor to the search for explanation, but it is not the only motivation. A reasonable version of the familiarity theory has more going for it than many of its critics suppose, but does not by itself provide an adequate description of our explanatory practices.

The third and best known account of explanation is the deductive-nomological model, according to which we explain a phenomenon by deducing it from a set of premises that includes at least one law that is necessary for the deduction (Hempel, 1965, pp. 335–76). On this view, we can explain why a particular star has its characteristic spectrum shifted towards the red by deducing this shift from the speed at which the star is receding from us, and the Doppler law that links recession and red shift. (This Doppler effect is similar to the change in pitch of a train whistle as the train passes by.) Of course, many scientific explanations, and most lay explanations, do not meet the strict requirements of the model, since they either do not contain exceptionless laws or do not strictly entail the phenomena, but they can be seen as 'explanation sketches' (Hempel, 1965, pp. 423–4) that provide better or worse approximations to a full deductive-nomological explanation.

The deductive-nomological model is closely related to the reason model, since the premises that enable us to deduce the phenomenon often also provide us with a reason to believe that the phenomenon occurs, but it avoids some of the weaknesses of the reason model. The explanation of the red shift in terms of the recession satisfies the deductive-nomological model even though, as it happens, the red shift is itself crucial evidence for the speed of recession that explains it. That is, the Doppler explanation is self-evidencing. Unlike the reason model, the deductive-nomological model allows for self-evidencing explanations. It also does better than the reason model in accounting for the difference between knowing and understanding. When we know that a phenomenon occurs but do not understand why, we usually do not know laws and supporting premises that entail the phenomenon. So at least a deductive-nomological argument usually gives us something new. To its credit, it also does seem that, when theories are used to give scientific explanations, these explanations often do aspire to deductive-nomological form. Moreover, the model avoids the main objection to the familiarity model, since a phenomenon may be common and unsurprising, but still await a deductive-nomological explanation.

The model nevertheless faces debilitating objections. It is almost certainly too strong: very few explanations fully meet the requirements of the model and, while some scientific explanations at least aspire to deductive-nomological status, many ordinary explanations include no laws and allow no deduction, yet are not incomplete or mere sketches. The model is also too weak. Perhaps the best known objection is that it does not account for the asymmetries of explanation (Bromberger, 1966; Friedman, 1974, p. 8; van Fraassen, 1980, p. 112). Consider again the Doppler explanation of the red shift. In that explanation, the law is used to deduce the shift from the recession, but the law is such that we could equally well use it to deduce the recession from the shift. Indeed this is how we figure out what the recession is. What follows from this is that, according to the deductive-nomological model, we can explain why the star is receding as it is by appeal to its red shift. But this is wrong: the shift no more explains the recession than a wet sidewalk explains why it is raining. The model does not account for the many cases where there is explanatory asymmetry but deductive symmetry.

The deductive-nomological model should produce a sense of *déjà vu*, since it is isomorphic to the hypothetico-deductive model of confirmation, which we considered in the last chapter. In one case we explain a phenomenon by deducing it from a law; in the other we show that the evidence confirms a hypothesis by deducing the evidence from the hypothesis. That deduction should play a role both in explanation and in inductive support is not itself suspicious, but the isomorphism of the models suggests that weaknesses of one may also count against the other, and this turns out to be the case. As we saw, the main weakness of the hypothetico-deductive model is that it is over-permissive, counting almost any datum as evidence for almost any hypothesis. The deductive-nomological model similarly makes it far too easy to explain (Hempel, 1965, p. 273, n. 33; pp. 293–4). This comes out most clearly if we consider the explanation of a general phenomenon, which is itself described by a law. Suppose, for example, that we wish to explain why the planets move in ellipses. According to the deductive-nomological model, we can 'explain' the ellipse law by deducing it from the conjunction of itself and any law you please, say a law in economics. The model also suffers from a problem analogous to the raven paradox. We may explain why an object warmed by pointing out that it was in the sun and everything warms when in the sun, but we cannot explain why an object was not in the sun by pointing out that it was not warmed. Just as the hypothetico-deductive model leaves the class of hypotheses confirmed by the available data dramatically underdetermined, so the deductive-nomological model underdetermines the class of acceptable explanations for a given phenomenon.

I conclude that none of the three models we have briefly considered gives an adequate description of our explanatory practices. Each of them captures distinctive features of certain explanations: some explanations provide reasons for belief, some make the initially unfamiliar or surprising familiar, and some are deductions from laws. But there are explanations that have none of these features, and something can have all of these features without being an acceptable explanation. I want now to consider a fourth model of explanation, the causal model. This model has its share of difficulties, but I believe it is more promising than the other three. I also can offer a modest development of the model that improves its descriptive adequacy and that will make it an essential tool for the discussion of Inference to the Best Explanation to follow. For these reasons, the causal model deserves a chapter of its own.

31

3

THE CAUSAL MODEL

FACT AND FOIL

According to the causal model of explanation, to explain a phenomenon is simply to give information about its causal history (cf. Lewis, 1986) or, where the phenomenon is itself a causal regularity, to explain it is to give information about the mechanism linking cause and effect. If we explain why smoking causes cancer, we do not give a cause of this causal connection, but we do give information about the causal mechanism that makes it. Not only is the causal model of explanation natural and plausible, but it seems to avoid many of the problems that beset the other views. It provides a clear distinction between understanding why a phenomenon occurs and merely knowing that it does; moreover, it does so in a way that makes understanding unmysterious and objective. Understanding is not some sort of super-knowledge, but simply more knowledge: knowledge of the phenomenon and knowledge about its causes. The model makes it clear how something can explain without itself being explained, and so avoids the regress of whys. One can know a phenomenon's cause without knowing the cause of that cause. Unlike the reason model, which requires that an explanation provide a reason to believe the phenomenon occurs, the causal model accounts for the legitimacy of self-evidencing explanations, where the phenomenon is an essential part of the evidence for the explanation. The causal model also avoids the most serious objection to the familiarity model, since a phenomenon can be common and unsurprising, even though we do not know its cause. Finally, it avoids many of the objections to the deductive-nomological model. Ordinary explanations do not

32

have to meet the requirements of that model, because one need not give a law to give a cause, and one need not know a law to have good reason to believe that a cause is a cause. As for the over-permissiveness of the deductive-nomological model, the reason recession explains red shift but not conversely is that causes explain effects and not conversely; the reason a conjunction does not explain its conjuncts is that conjunctions do not cause their conjuncts; and the reason the sun explains the warmth, while not being warmed does not explain not being in the sun, is that the sun causes an object to warm, but not being warmed does not cause something to be in the shade.

There are three natural objections to the causal model of explanation. The first is that we do not, as yet, have an uncontroversial analysis of causation. This, however, is no reason to abjure the model. The notion of causation is indispensable to philosophy, ordinary life, and much of science, and if we wait for a fully adequate analysis of causation before we use it to analyze other things, we may have to wait forever. I will not, in this book, say anything about what causation is, but trust that what I do say about the role of causation in explanation and inference holds for whatever the correct analysis of the notion turns out to be.

The second and perhaps the most obvious objection to the causal model of explanation is that there are non-causal explanations. Mathematicians and philosophers, for example, give explanations, but mathematical explanations are never causal, and philosophical explanations seldom are. A mathematician may explain why Gödel's Theorem is true, and a philosopher may explain why there can be no inductive justification of induction, but these are not explanations that cite causes. (Some philosophical explanations are, however, broadly causal, such as the explanations of inferential and explanatory practices that we are considering in this book.) In addition to the mathematical and philosophical cases, there are explanations of the physical world that seem non-causal. Here is a personal favorite. Suppose that some sticks are thrown into the air with a lot of spin, so that they separate and tumble about as they fall. Now freeze the scene at some point during the sticks' descent. Why are appreciably more of them near the horizontal axis than near the vertical, rather than in more or less equal numbers near each orientation, as one would have expected? The answer, roughly speaking, is that there are many more ways for a stick to be near the horizontal than near the vertical. To see this, consider purely horizontal and vertical orientations for a single stick with a fixed midpoint. There are infinitely many of the former, but only two of the latter. Or think of the shell that the ends of that stick trace as it takes every possible orientation. The areas that correspond to near the vertical are caps centered on the north and south poles formed when the stick is forty-five degrees or less off the vertical, and this area is substantially less than half the surface area of the entire sphere. Less roughly, the explanation why more sticks are near the vertical than near the horizontal is that there are two horizontal dimensions but only one vertical one. This is a lovely explanation, but apparently not a causal one, since geometrical facts cannot be causes. Non-causal explanations show that a causal model of explanation cannot be complete. Nevertheless, I believe that the causal view is still our best bet, because of the backward state of alternate views of explanation, and the overwhelming preponderance of causal explanations among all explanations. Nor does it seem *ad hoc* to limit our attention to causal explanations. The causal view does not simply pick out a feature that certain explanations happen to have: causal explanations are explanatory *because* they are causal. The analogous claims cannot, I think, be truly made for the other three models we have considered.

The third objection is that the causal model is too weak, that it underdetermines our explanatory practices. Let us focus on the causal explanation of particular events. We may explain an event by giving some information about its causal history, but causal histories are long and wide, and most causal information does not provide a good explanation. The big bang is part of the causal history of every event, but explains only a few. The spark and the oxygen are both part of the causal history that led up to the fire, but only one of them explains it. In a particular context, most information about the causal history of a phenomenon is explanatorily irrelevant, so explaining cannot simply be giving such information. This is an important objection, but I prefer to see it as a challenge. How can the causal model be developed to account for the causal selectivity of our explanatory practices? The rest of this chapter is a partial answer to this question. The answer is interesting in its own right, and it will also turn out to be a crucial tool for the job of developing and assessing an account of inference as Inference to the Best Explanation, the job that will be the focus of the balance of this book.

What makes one piece of information about the causal history of an event explanatory and another not? The short answer is that the causes that explain depend on our interests. But this does not yield a very informative model of explanation unless we can go some way towards spelling out how explanatory interests determine explanatory causes. One natural way to show how interests help us to select from among causes is to reveal additional structure in the phenomenon to be explained, structure that varies with interest and that points to particular causes. The idea here is that we can account for the specificity of explanatory answers by revealing the specificity in the explanatory question, where a difference in interest is an interest in explaining different things. Suppose we started by construing a phenomenon to be explained simply as a concrete event, say a particular eclipse. The number of causal factors is enormous. As Hempel has observed, however, we don't explain events, only aspects of events (Hempel, 1965, pp. 421–3). We don't explain the eclipse *tout court*, but only why it lasted as long as it did, or why it was partial, or why it was not visible from a certain place. Which aspect we ask about depends on our interests, and reduces the number of causal factors we need to consider for any particular phenomenon, since there will be many causes of the eclipse that are not, for example, causes of its duration. More recently, it has been argued that the interest relativity of explanation can be accounted for with a contrastive analysis of the phenomenon to be explained. What gets explained is not simply 'Why this?', but 'Why this *rather than* that?' (Garfinkel, 1981, pp. 28–41; van Fraassen, 1980, pp. 126–9). A contrastive phenomenon consists of a fact and a foil, and the same fact may have several different foils. We may not explain why the leaves turn yellow in November, but only, for example, why they turn yellow in November rather than in January, or why they turn yellow in November rather than turning blue.

The contrastive analysis of explanation is extremely natural. We often pose our why-questions explicitly in contrastive form and it is not difficult to come up with examples where different people select different foils, requiring different explanations. When I asked my 3-year-old son why he threw his food on the floor, he told me that he was full. This may explain why he threw it on the floor rather than eating it, but I wanted to know why he threw it rather than leaving it on his plate. An explanation of why I went to see *Jumpers* rather than *Candide* will probably not explain why I went to see *Jumpers* rather than staying at home, an explanation of why Able rather than Baker got the philosophy job may not explain why Able rather than Charles got the job, and an explanation of why the mercury in a thermometer rose rather than fell may not explain why it rose rather than breaking the glass. The proposal that phenomena to be explained have a complex fact–foil structure can be seen as another step along Hempel's path of focusing explanation by adding structure to the why-question. A fact is often not specific enough: we also need to specify a foil. Since the causes that explain a fact relative to one foil will not generally explain it relative to another, the contrastive question provides a further restriction on explanatory causes.

The role of contrasts in explanation will not account for all the factors that determine which cause is explanatory. For one thing, I do not assume that all why-questions are contrastive. For another, even in the cases of contrastive questions, the choice of foil is not, as we will see, the only relevant factor. Nevertheless, it does provide a central mechanism, so I want to show in some detail how contrastive questions help select explanatory causes. My discussion will fall into three parts. First, I will make three general observations about contrastive explanation. Then, I will use these observations to show why contrastive questions resist reduction to non-contrastive form. Finally, I will describe the mechanism of 'causal triangulation' by which the choice of foils in contrastive questions helps to select explanatory causes.

When we ask a contrastive why-question – 'Why the fact rather than the foil?' – we presuppose that the fact occurred and that the foil did not. Often we also suppose that the fact and the foil are in some sense incompatible. When we ask why Kate rather than Frank won the Philosophy Department Prize, we suppose that they could not both have won. Similarly, when we asked about leaves, we supposed that if they turn yellow in November, they cannot turn yellow in January, and if they turn yellow in November they cannot also turn blue. Indeed, it is widely supposed that fact and foil are always incompatible (Garfinkel, 1981, p. 40; Ruben 1987; Temple, 1988, p. 144). My first observation is that this is false: many contrasts are compatible. We often ask a contrastive question when we do not understand why two apparently similar situations turned out differently. In such a case, far from supposing any incompatibility between fact and foil, we ask the question just because we expected them to turn out the same. By the time we ask the question, we realize that our expectation was disappointed, but this does not normally lead us to believe that the fact precluded the foil, and the explanation for the contrast will usually not show that it did. Consider the much discussed example of syphilis and paresis (cf. Hempel, 1965, pp. 369–70; van Fraassen, 1980, p. 128). Few with syphilis contract paresis, but we can still explain why Jones rather than Smith contracted paresis by pointing out that only Jones had syphilis. In this case, there is no incompatibility. Only Jones contracted paresis, but they both could have: Jones's affliction did not protect Smith. Of course, not every pair of compatible fact and foil would yield a sensible why-question but, as we will see, it is not necessary to restrict contrastive why-questions to incompatible contrasts to distinguish sensible questions from silly ones.

My second and third observations concern the relationship between an explanation of the contrast between a fact and foil and the explanation of the fact alone. I do not have a general account of what it takes to explain a fact on its own. As we will see, this is not necessary to give an account of what it takes to explain a contrast; indeed, this is one of the advantages of a contrastive analysis. Yet, based on our intuitive judgments of what is and what is not an acceptable explanation of a fact alone, we can see that the requirements for explaining a fact are different from the requirements for explaining a contrast. My second observation, then, is that explaining a contrast is sometimes easier than explaining the fact alone (cf. Garfinkel, 1981, p. 30). An explanation of 'P rather than Q' is not always an explanation of P. This is particularly clear in examples of compatible contrasts. Jones's syphilis does not explain why he got paresis, since the vast majority of people who get syphilis do not get paresis, but it does explain why Jones rather than Smith got paresis, since Smith did not have syphilis. The relative ease with which we explain some contrasts also applies to many cases where there is an incompatibility between fact and foil. My preference for contemporary plays may not explain why I went to see Jumpers last night, since it does not explain why I went out, but it does explain why I went to see Jumpers rather than Candide.

A particularly striking example of the relative ease with which some contrasts can be explained is the explanation that I chose A rather than B because I did not realize that B was an option. If you ask me why I ordered eggplant rather than sea bass (a 'daily special'), I may give the perfectly good answer that I did not know there were any specials; but this would not be an acceptable answer to the simple question, 'Why did you order eggplant?'. One reason we can sometimes explain a contrast without explaining the fact alone seems to be that contrastive questions incorporate a presupposition that makes explanation easier. To explain 'P rather than Q' is to give a certain type of explanation of P, *given* 'P or Q', and an explanation that succeeds with the presupposition will not generally succeed without it.

My final observation is that explaining a contrast is also sometimes harder than explaining the fact alone. An explanation of P is not always an explanation of 'P rather than Q'. This is obvious in the case of compatible contrasts: we cannot explain why Jones rather than Smith contracted paresis without saying something about Smith. But it also applies to incompatible contrasts. To explain why I went to *Jumpers* rather than *Candide*, it is not enough for me to say that I was in the mood for a philosophical play. To explain why Kate rather than Frank won the prize, it is not enough that she wrote a good essay; it must have been better than Frank's. One reason why explaining a contrast is sometimes harder than explaining the fact alone is that explaining a contrast requires giving causal information to distinguish the fact from the foil, and information that we accept as an explanation of the fact alone may not do this.

FAILED REDUCTIONS

There have been a number of attempts to reduce contrastive questions to non-contrastive and generally truth-functional form. One motivation for this is to bring contrastive explanations into the fold of the deductive-nomological model since, without some reduction, it is not clear what the conclusion of a deductive explanation of 'P rather than Q' ought to be. Armed with our three observations – that contrasts may be compatible, and that explaining a contrast is sometimes easier and sometimes harder than explaining the fact alone – we can show that contrastive questions resist a reduction to non-contrastive form. We have already seen that the contrastive question 'Why P rather than Q?' is not equivalent to the simple question 'Why P?', where two why-questions are explanatorily equivalent just in case any adequate answer to one is an adequate answer to the other. One of the questions may be easier or harder to answer than the other. Still, a proponent of the deductive-nomological model of explanation may be tempted to say that, for incompatible contrasts, the question 'Why P rather than Q?' is equivalent to 'Why P?'. But it is not plausible to say that a deductive-nomological explanation of P is generally necessary to explain 'P rather than Q'. More interestingly, a deductive-nomological explanation of P is not always sufficient to explain 'P rather than Q', for any incompatible Q. Imagine a typical deductive explanation for the rise of mercury in a thermometer. Such an explanation would explain various contrasts, for example why the mercury rose rather than fell. It may not, however, explain why the mercury rose rather than breaking the glass. A full deductive-nomological explanation of the rise will have to include a premise saying that the glass does not break, but it does not need to explain this.

Another natural suggestion is that the contrastive question 'Why P rather than Q?' is equivalent to the conjunctive question 'Why P and not-Q?' On this view, explaining a contrast between fact and foil is tantamount to explaining the conjunction of the fact and the negation of the foil (Temple, 1988). In ordinary language, a contrastive question is often equivalent to its corresponding conjunction, simply because the 'and not' construction is often used contrastively. Instead of asking, 'Why was the prize won by Kate rather than by Frank?', the same question could be posed by asking 'Why was the prize won by Kate and not by Frank?' But this colloquial equivalence does not seem to capture the point of the conjunctive view. To do so, I suggest that the conjunctive view be taken to entail that explaining a conjunction at least requires explaining each conjunct; that an explanation of 'P and not-Q' must also provide an explanation of P and an explanation of not-Q. Thus, on the conjunctive view, to explain why Kate rather than Frank won the prize at least requires an explanation of why Kate won it and an explanation of why Frank did not. This account of contrastive explanation falls to the observation that explaining a contrast is sometimes easier than explaining the fact alone, since explaining P and explaining not-Q is at least as difficult as explaining P.

39

The observations that explaining contrasts is sometimes easier and sometimes harder than explaining the fact alone reveal another objection to the conjunctive view, on any model of explanation that is deductively closed. A model is deductively closed if it entails that an explanation of P will also explain any logical consequence of P. Consider cases where the fact is logically incompatible with the foil. Here P entails not-Q, so the conjunction 'P and not-Q' is logically equivalent to P alone. Furthermore, all conjunctions whose first conjunct is P and whose second conjunct is logically incompatible with P will be equivalent to each other, since they are all logically equivalent to P. Hence, for a deductively closed model of explanation, explaining 'P and not-Q' is tantamount to explaining P, whatever Q may be, so long as it is incompatible with P. We have seen, however, that explaining 'P rather than Q' is not generally tantamount to explaining P. The conjunction is explanatorily equivalent to P, and the contrast is not, so the conjunction is not equivalent to the contrast.

The failure to represent a contrastive phenomenon by the fact alone or by the conjunction of the fact and the negation of the foil suggests that, if we want a non-contrastive paraphrase, we ought instead to try something logically weaker than the fact. In some cases it does seem that an explanation of the contrast is really an explanation of a logical consequence of the fact. This is closely related to what Hempel has to say about 'partial explanation' (1965, pp. 415–18). He gives the example of Freud's explanation of a particular slip of the pen that resulted in writing down the wrong date. Freud explains the slip with his theory of wish-fulfillment, but Hempel objects that the explanation does not really show why that particular slip took place, but at best only why there was some wish-fulfilling slip or other. Freud gave a partial explanation of the particular slip, since he gave a full explanation of the weaker claim that there was some slip. Hempel's point fits naturally into contrastive language: Freud did not explain why it was this slip rather than another wish-fulfilling slip, though he did explain why it was this slip rather than no slip at all. And it seems natural to analyze 'Why this slip rather than no slip at all?' as 'Why some slip?' In general, however, we cannot paraphrase contrastive questions with consequences of their facts. We cannot, for example, say that to explain why the leaves turn yellow in November rather than in

January is just to explain why the leaves turn (some color or other) in November. This attempted paraphrase fails to discriminate between the intended contrastive question and the question, 'Why do the leaves turn in November rather than fall right off?' Similarly, we cannot capture the question, 'Why did Jones rather than Smith get paresis?', by asking about some consequence of Jones's condition, such as why he contracted a disease.

A general problem with finding a paraphrase entailed by the fact P is that, as we have seen, explaining a contrast is sometimes harder than explaining P alone. There are also problems peculiar to the obvious candidates. The disjunction, 'P or Q' will not do: explaining why I went to *Jumpers* rather than *Candide* is not the same as explaining why I went to either. Indeed, this proposal gets things almost backwards: the disjunction is what the contrastive question assumes, not what calls for explanation. This suggests, instead, that the contrast is equivalent to the conditional, 'if P or Q, then P' or, what comes to the same thing if the conditional is truth-functional, to explaining P on the assumption of 'P or Q'. Of all the reductions we have considered, this proposal is the most promising, but I do not think it will do. On a deductive model of explanation it would entail that any explanation of not-Q is also an explanation of the contrast, which is incorrect. We cannot explain why Jones rather than Smith has paresis by explaining why Smith did not get it. It would also wrongly entail that any explanation of P is an explanation of the contrast, since P entails the conditional.

CAUSAL TRIANGULATION

By asking a contrastive question, we can achieve a specificity that we do not seem to be able to capture either with a non-contrastive sentence that entails the fact or with one that the fact entails. But how then does a contrastive question specify the sort of information that will provide an adequate answer? It now appears that looking for a non-contrastive reduction of 'P rather than Q' is not a useful way to proceed. The contrastive claim may entail no more than 'P and not-Q' or perhaps better, 'P but not-Q', but explaining the contrast is not the same as explaining these conjuncts. We will do better to leave the analysis of the contrastive question to one side, and instead consider directly what it takes

41

to provide an adequate answer. David Lewis has given an interesting account of contrastive explanation that does not depend on paraphrasing the contrastive question. According to him, we explain why event P occurred rather than event Q by giving information about the causal history of P that would not have applied to the history of Q, if Q had occurred (Lewis, 1986, pp. 229–30). Roughly, we cite a cause of P that would not have been a cause of Q. In Lewis's example, we can explain why he went to Monash rather than to Oxford in 1979 by pointing out that only Monash invited him, because the invitation to Monash was a cause of his trip, and that invitation would not have been a cause of a trip to Oxford, if he had taken one. On the other hand, Lewis's desire to go to places where he has good friends would not explain why he went to Monash rather than Oxford, since he has friends in both places and so the desire would have been part of either causal history.

Lewis's account, however, is too weak: it allows for unexplanatory causes. Suppose that both Oxford and Monash had invited him, but he went to Monash anyway. On Lewis's account, we can still explain this by pointing out that Monash invited him, since *that* invitation still would not have been a cause of a trip to Oxford. Yet the fact that he received an invitation from Monash clearly does not explain why he went there rather than to Oxford in this case, since Oxford invited him too. Similarly, Jones's syphilis satisfies Lewis's requirement even if Smith has syphilis too, since Jones's syphilis would not have been a cause of Smith's paresis, had Smith contracted paresis, yet in this case Jones's syphilis would not explain why he rather than Smith contracted paresis.

It might be thought that Lewis's account could be saved by construing the causes more broadly, as types rather than tokens. In the case of the trip to Monash, we might take the cause to be receiving an invitation rather than the particular invitation to Monash he received. If we do this, we can correctly rule out the attempt to explain the trip by appeal to an invitation if Oxford also invited since, in this case, receiving an invitation would also have been a cause of going to Oxford. This, however, will not do, for two reasons. First, it does not capture Lewis's intent: he is interested in particular elements of a particular causal history, not general causal features. Second, and more importantly, the suggestion throws out the baby with the bath water. Now we have also ruled out the perfectly good explanation by invitation in some cases where only Monash invites. To see this, suppose that Lewis is the sort of person who only goes where he is invited. In this case, an invitation would have been part of a trip to Oxford, if he had gone there.

To improve on Lewis's account, consider John Stuart Mill's Method of Difference, his version of the controlled experiment, which we discussed in chapter one (Mill, 1904, III.VIII.2). Mill's Method rests on the principle that a cause must lie among the antecedent differences between a case where the effect occurs and an otherwise similar case where it does not. The difference in effect points back to a difference that locates a cause. Thus we might infer that contracting syphilis is a cause of paresis, since it is one of the ways Smith and Jones differed. The cause that the Method of Difference isolates depends on which control we use. If, instead of Smith, we have Doe, who does not have paresis but did contract syphilis and had it treated, we would be led to say that a cause of paresis is not syphilis, but the failure to treat it. The Method of Difference also applies to incompatible as well as to compatible contrasts. As Mill observes, the Method often works particularly well with diachronic (before and after) contrasts, since these give us histories of fact and foil that are largely shared, making it easier to isolate a difference. If we want to determine the cause of a person's death, we naturally ask why he died when he did rather than at another time, and this yields an incompatible contrast, since you can only die once.

The Method of Difference concerns the discovery of causes rather than the explanation of effects, but the similarity to contrastive explanation is striking (cf. Garfinkel 1981, p. 40). Accordingly, I propose that, for the causal explanations of events, explanatory contrasts select causes by means of what I will call the 'Difference Condition'. *To explain why P rather than Q, we must cite a causal difference between P and not-Q, consisting of a cause of P and the absence of a corresponding event in the history of not-Q.* Instead of pointing to a counterfactual difference, a particular cause of P that would not have been a cause of Q, as Lewis suggests, contrastive questions select as explanatory an actual difference between P and not-Q. Lewis's invitation to Monash does not explain why he went there rather than to Oxford if he was invited to both places because, while there is an invitation in the history of his trip to Monash, there is also an invitation in the history that led him to forgo Oxford. Similarly, the Difference Condition correctly entails that Jones's syphilis does not explain why he rather than Smith contracted paresis if Smith had syphilis too, and that Kate's submitting an essay does not explain why she rather than Frank won the prize. Consider now some of the examples of successful contrastive explanation. If only Jones had syphilis, that explains why he rather than Smith has paresis, since having syphilis is a condition whose presence was a cause of Jones's paresis and a condition that does not appear in Smith's medical history. Writing the best essay explains why Kate rather than Frank won the prize, since that is a causal difference between the two of them. Lastly, the fact that *Jumpers* is a contemporary play and *Candide* is not caused me both to go to one and to avoid the other.

The application of the Difference Condition is easiest to see in cases of compatible contrasts, since here the causal histories of P and of not-Q are generally distinct, but the condition does not require this. In cases of choice, for example, the causal histories are usually the same: the causes of my going to *Jumpers* are the same as the causes of my not going to *Candide*. The Difference Condition may nevertheless be satisfied if my belief that *Jumpers* is a contemporary play is a cause of going, and I do not believe that *Candide* is a contemporary play. That is why my preference for contemporary plays explains my choice. Similarly, the invitation from Monash explains why Lewis went there rather than to Oxford and satisfies the Difference Condition, so long as Oxford did not invite. The condition does not require that the same event be present in the history of P but absent in the history of not-Q, a condition that could never be satisfied when the two histories are the same, but only that the cited cause of P find no corresponding event in the history of not-Q where, roughly speaking, a corresponding event is something that would bear the same relation to Q as the cause of P bears to P.

One of the merits of the Difference Condition is that it brings out the way the incompatibility of fact and foil, when it obtains, is not sufficient to transform an explanation of the fact into an explanation of the contrast, even if the cause of the fact is also a cause of the foil not obtaining. Perhaps we could explain why Able got the philosophy job by pointing out that Quine wrote him a strong letter of recommendation, but this will only explain why Able rather than Baker got the job if Quine did not also write a

44

similar letter for Baker. If he did, Quine's letter for Able does not alone explain the contrast, even though that letter is a cause of both Able's success and Baker's failure, and the former entails the latter. The letter may be a partial explanation of why Able got the job, but it does not explain why Able rather than Baker got the job. In the case where they both have strong letters from Quine, a good explanation of the contrast will have to find an actual difference, say that Baker's dossier was weaker than Able's in some other respect, or that Able's specialties were more useful to the department. There are some cases of contrastive explanation that do seem to rely on the way the fact precludes the foil, but I think these can be handled by the Difference Condition. For example, suppose we explain why a bomb went off prematurely at noon rather than in the evening by saying that the door hooked up to the trigger was opened at noon (I owe this example to Eddy Zemach). Here it may appear that the Difference Condition is not in play, since the explanation would stand even if the door was also opened in the evening. But the Difference Condition is met, if we take the cause not simply to be the opening of the door, but the opening of the door when it is rigged to an armed bomb.

My goal in this chapter is to show how the choice of contrast helps to determine an explanatory cause, not to show why we choose one contrast rather than another. The latter question is not part of providing a model of explanation, as that task has been traditionally construed. It is no criticism of the deductive-nomological model that it does not tell us which phenomena we care to explain, so long as it tells us what counts as an adequate explanation of the phenomena we select; similarly, it is no criticism of my account of contrastive explanation that it does not tell us why we are interested in explaining some contrasts rather than others. Still, an account of the considerations that govern our choice of why-questions ought to form a part of a full model of our explanatory practices, and it is to the credit of the contrastive analysis that it lends itself to this. As we will see in later chapters, our choice of why-questions is often governed by our *inferential* interests, so that we choose contrasts that help us to determine which of competing explanatory hypotheses is correct. For now, however, we may just note that not all contrasts make for sensible contrastive questions. It does not make sense, for example, to ask why Lewis went to Monash rather than Baker getting the philosophy job. One might have thought that a sensible contrast

45

must be one where fact and foil are incompatible, but we have seen that this is not necessary, since there are many sensible compatible contrasts. There are also incompatible contrasts that do not yield reasonable contrastive questions, such as why someone died when she did rather than never having been born. The Difference Condition suggests instead that the central requirement for a sensible contrastive question is that the fact and the foil have a largely similar history, against which the differences stand out. When the histories are too disparate, we do not know where to begin to answer the question. There are, of course, other considerations that help to determine the contrasts we actually choose. For example, in the case of incompatible contrasts, we often pick as foil the outcome we expected; in the case of compatible contrasts, as I have already mentioned, we often pick as foil a case we expected to turn out the same way as the fact. The condition of a similar history also helps to determine what will count as a corresponding event, a notion that my general account does not make precise. If we were to ask why Lewis went to Monash rather than Baker getting the job, it would be difficult to see what in the history of Baker's failure would correspond to Lewis's invitation, but when we ask why Able rather than Baker got the job, the notion of a corresponding event is relatively clear.

I will now consider three further issues connected with my analysis of contrastive explanation: the need for further principles for distinguishing explanatory from unexplanatory causes, the prospects for treating all why-questions as contrastive, and a more detailed comparison of my analysis with the deductive-nomological model. When we ask contrastive why-questions, we choose our foils to point towards the sort of causes that interest us. As we have just seen, when we ask about a surprising event, we often make the foil the thing we expected. This focuses our inquiry on causes that will illuminate the reason our expectation went wrong. Failed expectations are not, however, the only things that prompt us to ask why-questions. If a doctor is interested in the internal etiology of a disease, he will ask why the afflicted have it rather than other people in similar circumstances, even though the shared circumstances may be causally relevant to the disease. Again, if a machine malfunctions, the natural diagnostic contrast is its correct behavior, since that directs our attention to the causes that we want to change. But the contrasts we construct will almost always leave multiple differences that meet the Difference Condition. More than one of these may be explanatory: my account does not entail that there is only one way to explain a contrast. At the same time, however, some causally relevant differences will not be explanatory in a particular context, so while the Difference Condition may be necessary for the causal contrastive explanations of particular events, it is not generally sufficient. For that we need further principles of causal selection.

The considerations that govern selection from among causally relevant differences are numerous and diverse; the best I can do here is to mention what a few of them are. An obvious pragmatic consideration is that someone who asks a contrastive question may already know about some causal differences, in which case a good explanation will have to tell her something new. If she asks why Kate rather than Frank won the prize, she may assume that it was because Kate wrote the better essay, in which case we will have to tell her more about the differences between the essays that made Kate's better. A second consideration is that, when they are available, we usually prefer explanations where the foil would have occurred if the corresponding cause had occurred. Suppose that only Able had a letter from Quine, but even a strong letter from Quine would not have helped Baker much, since his specialties do not fit the department's needs. Suppose also that, had Baker's specialties been appropriate, he would have gotten the job, even without a letter from Quine. In this case, the difference in specialties is a better explanation than the difference in letters. Note, however, that an explanation that does not meet this condition of counterfactual sufficiency for the occurrence of the foil may be perfectly acceptable, if we do not know of a sufficient difference. The explanation of why Jones rather than Smith contracted paresis is an example of this: even if Smith had syphilis in his medical history, he probably would not have contracted paresis. Moreover, even in cases where a set of known causes does supply a counterfactually sufficient condition, the inquirer may be much more interested in some than in others. The doctor may be particularly interested in causes he can control, the lawyer in causes that are connected with legal liability, and the accused in causes that cannot be held against him.

We also prefer differences where the cause is causally necessary for the fact in the circumstances. Consider a case of over-

47

determination. Suppose that you ask me why I ordered eggplant rather than beef, when I was in the mood for eggplant and not for beef, and I am a vegetarian. My mood and my convictions are separate causes of my choice, each causally sufficient in the circumstance and neither necessary. In this case, it would be better to give both differences than just one. The Difference Condition could easily be modified to require necessary causes, but I think this would make the condition too strong. One problem would be cases of 'failsafe' overdetermination. Suppose we change the restaurant example so that my vegetarian convictions were not a cause of the particular choice I made: that time, it was simply my mood that was relevant. Nevertheless, even if I had been in the mood for beef, I would have resisted, because of my convictions. In this case, my explanation does not have to include my convictions, even though my mood was not a necessary cause of my choice. (Of course if I knew that you were asking me about my choice because you were planning to invite me to your house for dinner, it would be misleading for me not to mention my convictions, but this goes beyond the conditions for explaining the particular choice I made.) Again, we sometimes don't know whether a cause is necessary for the effect, and in such cases the cause still seems explanatory. But when there are differences that supply a necessary cause, and we know that they do, we prefer them.

Another reason why satisfying the Difference Condition is not always sufficient for a good contrastive explanation is that a difference favoring the fact may be balanced against another favoring the foil. If I tell you that Lewis went to Monash rather than Oxford because only Monash invited him, you might reply, 'Yes, but Oxford has much better bookshops, and Lewis loves bookshops.' In such a case, I will have to supplement my original explanation by showing, or at least claiming, that the actual cause of the foil 'trumped' the potential cause of the foil. Thus I might claim that his preference for places that invite him was stronger than his preference for places with outstanding bookshops. Of course this might not be true: the difference I originally cite may not by itself be stronger than the countervailing force you mention. In this case, I must find other or additional differences that are. There are doubtless other principles that also play a role in determining which differences yield the best explanation in a particular context. So there is more to contrastive explanation than the Difference Condition describes, but that condition does seem to describe the central mechanism of causal selection.

Since contrastive questions are so common and foils play such an important role in determining explanatory causes, it is natural to wonder whether all why-questions are not at least implicitly contrastive. Often the contrast is so obvious that it isn't worth mentioning. If you ask me why I was late for our appointment, the question is why I was late rather than on time, not why I was late rather than not showing up at all. Moreover, in cases where there is no specific contrast, stated or implied, we might construe 'Why P?' as 'Why P rather than not-P?', thus subsuming all causal why-questions under the contrastive analysis. But the Difference Condition seems to misbehave for these 'global' contrasts. It requires that we find a cause of P that finds no corresponding cause in the history of not-Q. But, if the foil is simply the negation of the fact, this seems to require that we find a cause of P that finds no corresponding cause of itself, which is impossible, since it is tantamount to the requirement that we find a cause of P that is absent from the history of P.

We may, however, be able to analyze the explanation of P *simpliciter* as the explanation of P rather than not-P. The correct way to construe the Difference Condition as it applies to the limiting case of the contrast, P rather than not-P, is that we must find a difference for events logically or causally incompatible with P, not for a single event, 'not-P'. Suppose that we ask why Jones has paresis, with no implied contrast. This would require a difference for foils where he does not have paresis. Saying that he had syphilis differentiates between the fact and the foil of a thoroughly healthy Jones, but this is not enough, since it does not differentiate between the fact and the foil of Jones with syphilis but without paresis. Excluding many incompatible foils will push us towards a sufficient cause of Jones's syphilis, since it is only by giving such a 'full cause' that we can be sure that some bit of it will be missing from the history of all the foils. To explain P rather than not-P, however, we do not need to explain every incompatible contrast. We do not, for example, need to explain why Jones contracted paresis rather than being long dead or never being born. The most we can require is that we exclude all incompatible foils with histories similar to the history of the fact. One difficulty for this way of avoiding the pathological requirement of finding a cause of P that is absent from the history

49

of P is that there appear to be some facts whose negation also seems to be a single fact (I owe this point to Elliot Sober). Suppose we wish to understand why there are tigers. Here the foil seems simply to be the absence of tigers, and we cannot give a cause of the existence of tigers that is not in the history of tigers. But the existence of tigers is not an event, so this example does not affect my account, which is only meant to apply to the explanation of events. So perhaps the problem does not arise for contrasts whose facts are events and whose foils are either events or sets of events. The Difference Condition will apply to some contrasts that are not explicitly event-contrasts, but not to all of them. Even for P's that are events, however, I am not sure that every apparently non-contrastive question should be analyzed in contrastive form, so I am an agnostic about the existence of non-contrastive why-questions.

Finally, let us compare my analysis of contrastive explanation to the deductive-nomological model. First, as we have already seen, a causal view of explanation has the merit of avoiding all the counterexamples to the deductive-nomological model where causes are deduced from effects. It also avoids the unhappy consequence of counting almost every explanation we give as a mere sketch, since one can give a cause of P that meets the Difference Condition for various foils without having the laws and singular premises necessary to deduce P. Many explanations that the deductive model counts as only very partial explanations of P are in fact reasonably complete explanations of P rather than Q. The excessive demands of the deductive model are particularly striking for cases of compatible contrasts, at least if the deductive-nomologist requires that an explanation of P rather than Q provide an explanation of P and an explanation of not-Q. In this case, the model makes explaining the contrast substantially harder than providing a deductive explanation of P, when in fact it is often substantially easier. Our inability to find a non-contrastive reduction of contrastive questions is, among other things, a symptom of the inability of the deductive-nomological model to give an accurate account of this common type of explanation.

There are at least two other conspicuous advantages of a causal contrastive view of explanation over the deductive-nomological model. One odd feature of the model is that it entails that an explanation cannot be ruined by adding true premises, so long as the additional premises do not render the law superfluous to the deduction, by entailing the conclusion outright. This consequence follows from the elementary logical point that additional premises can never convert a valid argument into an invalid one. In fact, however, irrelevant additions can spoil an explanation. If I say that Jones rather than Smith contracted paresis because only Jones had syphilis and only Smith was a regular churchgoer, I have not simply said more than I need to, but I have given an incorrect explanation, since going to church is not a prophylactic. By requiring that explanatory information be causally relevant, the contrastive model avoids this problem. Another related and unhappy feature of the deductive-nomological model is that it entails that explanations are virtually deductively closed: an explanation of P will also be an explanation of any logical consequence of P, so long as the consequence is not directly entailed by the initial conditions alone. (For an example of the slight non-closure in the model, notice that a deductive-nomological explanation of P will not also be a deductive-nomological explanation of the disjunction of P and one of the initial conditions of the explanation.) In practice, however, explanation seems to involve a much stronger form of non-closure. I might explain why all the men in the restaurant are wearing paisley ties by appealing to the fashion of the times for ties to be paisley, but this might not explain why they are all wearing ties, which is because of a rule of the restaurant. (I owe this example to Tim Williamson.) The contrastive view gives a natural account of this sort of non-closure. When we ask about paisley ties, the implied foil is other sorts of tie; but when we ask simply about ties, the foil is not wearing ties. The fashion marks a difference in one case, but not in the other.

A defender of the deductive-nomological model may respond to some of these points by arguing that, whatever the merits of a contrastive analysis of lay explanation, the deductive model (perhaps with an additional restriction blocking 'explanations' of causes by effects) gives a better account of scientific explanation. For example, it has been claimed that scientific explanations, unlike ordinary explanations, do not exhibit interest relativity of foil variation that a contrastive analysis exploits, so a contrastive analysis does not apply to scientific explanation (Worrall, 1984, pp. 76–7). It is, however, a mistake to suppose that all scientific explanations even aspire to deductive-nomological status. The explanation of why Jones rather than Smith contracted paresis is presumably scientific, but it is not a deduction *manqué*. Moreover, as the example of the thermometer shows, even a full deductive-nomological explanation may exhibit interest relativity. I may explain the fact relative to some foils but not relative to others. A typical deductive-nomological explanation of the rise of mercury in a thermometer will simply assume that the glass does not break and so while it will explain, for example, why the mercury rose rather than fell, it will not explain why it rose rather than breaking the thermometer. Quite generally, a deductive-nomological explanation of a fact will not explain that fact relative to any foils that are themselves logically inconsistent with one of the premises of the explanation. Again, a Newtonian explanation of the earth's orbit (ignoring the influence of the other planets) will explain why the earth has its actual orbit rather than some other orbits, but it will not explain why the earth does not have any of the other orbits that are compatible with Newton's theory. The explanation must assume information about the earth's position and velocity at some time that will rule out the other Newtonian orbits, but it will not explain why the earth does not travel in those paths. To explain this would require quite different information about the early history of the earth. Similarly, an adaptionist explanation for a species' possession of a certain trait may explain why it has that trait rather than various maladaptive traits, but it may not explain why it had that trait rather than other traits that would perform the same functions equally well. To explain why an animal has one trait rather than another functionally equivalent trait requires instead appeal to the evolutionary history of the species, insofar as it can be explained at all.

With rather more justice, a deductive-nomologist might object that scientific explanations do very often essentially involve laws and theories, and that the contrastive view does not seem to account for this. For even if the fact to be explained carries no restricting contrast, the contrastive view, if it is extended to this case by analyzing 'Why P?' as 'Why P rather than not-P?', only requires at most that we cite a condition that is causally sufficient for the fact, not that we actually give any laws. In reply, one might mention first that laws may nevertheless be part of a correct analysis of the causal relation itself, and that knowledge of laws is sometimes essential evidence for causal claims. Moreover, the contrastive view can help to account for the explicit role of laws in many scientific explanations. To see this, notice that scientists are often and perhaps primarily interested in explaining regularities, rather than particular events (cf. Friedman, 1974, p. 5; though explaining particular events is also important when, for example, scientists test their theories, since observations are of particular events). I think that the Difference Condition applies to many explanations of regularities, but to give a contrastive explanation of a regularity will require citing a law, or at least a generalization, since here we need some general cause (cf. Lewis, 1986, pp. 225–6). To explain, say, why people feel the heat more when the humidity is high, we must find some general causal difference between cases where the humidity is high and cases where it is not, such as the fact that the evaporation of perspiration, upon which our cooling system depends, slows as the humidity rises. So the contrastive view, in an expanded version that applies to general facts as well as to events (a version I do not here provide), should be able to account for the role of laws in scientific explanations as a consequence of the scientific interest in general why-questions. Similarly, although the contrastive view does not require deduction for explanation, it is not mysterious that scientists should often look for explanations that do entail the phenomenon to be explained. This may not have to do with the requirements of explanation *per se*, but rather with the uses to which explanations are put. Scientists often want explanations that can be used for accurate prediction, and this requires deduction. Again, the construction of an explanation is a way to test a theory, and some tests require deduction.

Another way of seeing the compatibility of the scientific emphasis on theory and the contrastive view is by observing that scientists are not just interested in this or that explanation, but in a unified explanatory scheme. Scientists want theories, in part, because they want engines that will provide many explanations. The contrastive view does not entail that a theory is necessary for any particular explanation, but a good theory is the best way to provide the many and diverse contrastive explanations that the scientist is after. This also helps to account for the familiar point that scientists are often interested in discovering causal mechanisms. The contrastive view will not require a mechanism to explain why one input into a black box causes one output, but it pushes us to specify more and more of the detailed workings of the box as we try to explain its full behavior under diverse conditions. So I conclude that the contrastive view of explanation does not fly in the face of scientific practice.

The Difference Condition shows how contrastive questions about particular events help to determine an explanatory cause by a kind of *causal triangulation*. This contrastive model of causal explanation cannot be the whole story about explanation since, among other things, not all explanations are causal and since the choice of foil is not the only factor that affects the appropriate choice of cause. The model does, however, give a natural account of much of what is going on in many explanations, and it captures some of the merits of competing accounts while avoiding some of their weaknesses. We have just seen this in some detail for the case of the deductive-nomological model. It also applies to the familiarity model. When an event surprises us, a natural foil is the outcome we had expected, and meeting the Difference Condition for this contrast will help to show us why our expectation went wrong. The mechanism of causal triangulation also accounts for the way a change in foil can lead to a change in explanatory cause, since a difference for one foil will not in general be a difference for another. It also shows why explaining 'P rather than Q' is sometimes harder and sometimes easier than explaining P alone. It may be harder, because it requires the absence of a corresponding cause in the history of not-Q, and this is something that will not generally follow from the presence of the cause of P. Explaining the contrast may be easier, because the cause of P need not be sufficient for P, so long as it is part of a causal difference between P and not-Q. Causal triangulation also elucidates the interest relativity of explanation. We express some of our interests through our choice of foils and, by construing the phenomenon to be explained as a contrast rather than the fact alone, the interest relativity of explanations reduces to the important but unsurprising point that different people are interested in explaining different phenomena. Moreover, the Difference Condition shows that different interests do not require incompatible explanations to satisfy them, only different but compatible causes. The mechanism of causal triangulation also helps to account for the failure of various attempts to reduce contrastive questions to some non-contrastive form. None of these brings out the way a foil serves to select a location in the causal history leading up to the fact. Causal triangulation is the central feature of contrastive

54

explanation that non-contrastive paraphrases suppress. Lastly, we will find that the structure of contrastive explanations helps us with the problem of describing our inferential practices, a problem whose difficulties we met in chapter one, when it is wed to such a model of inference as Inference to the Best Explanation, a model to which we now finally turn.

4

INFERENCE TO THE BEST EXPLANATION

SPELLING OUT THE SLOGAN

Our initial survey of the problems of induction and explanation is now complete. We have considered some of the forms these problems take, some of the reasons they are so difficult to solve, and some of the weaknesses in various attempts to solve them. In the last chapter, I also attempted something more constructive, by giving what I hope is an improved version of the causal model of explanation. Up to now, however, we have treated inference and explanation virtually in mutual isolation, a separation that reflects most of the literature on these subjects. Although the discussion of inference in chapter one construed the task of describing our practices as itself an explanatory inquiry, the attempt to specify the black box mechanism that takes us from evidence to inference and so explains why we make the inferences we do, none of the models of inference we considered explicitly invoked explanatory relations between evidence and conclusion. Similarly, in the discussion of explanation in chapters two and three, inferential considerations played a role in only one of the models, the reason model. That model uses an inferential notion to account for explanation, by claiming that we explain a phenomenon by giving some reason to believe that the phenomenon occurs, and it was found to be unacceptable for inferential reasons, since it does not allow for self-evidencing explanations, such as the explanation of the tracks in the snow or of the red shift of the star, where the phenomenon that is explained provides an essential part of the reason for believing that the explanation is correct.

In this chapter, the relationship between the practices of explanation and of inference will take center stage, where it will remain for the rest of this book. Let us begin with a very simple view of that relationship. First we make our inferences; then, when we want to explain a phenomenon, we draw upon our pool of beliefs for an explanation, a pool filled primarily by those prior inferences. This, however, must be too simple, since our pool may not contain the explanation we seek. So a slightly less simple view is that, if we don't find an explanation in our pool, we search for a warranted inference that will explain, a process that may also require further observation. Explanatory considerations thus have some bearing on inference, since they may focus our inquiry but, on this view, inference still comes before explanation. After all, the most basic requirement of an explanation is that the explanatory information be correct, so how can we be in a position to use that information for an explanation unless we first know that it is correct?

This picture, however, seriously underestimates the role of explanatory considerations in inference. Those considerations tell us not only what to look for, but also whether we have found it. Take the cases of self-evidencing explanations. The tracks in the snow are the evidence for what explains them, that a person passed by on snowshoes; the red shift of the star is an essential part of the reason we believe the explanation, that it has a certain velocity of recession. In these cases, it is not simply that the phenomena to be explained provide reasons for inferring the explanations: we infer the explanations precisely because they would, if true, explain the phenomena. Of course, there is always more than one possible explanation for any phenomenon – the tracks might instead have been caused by a trained monkey on snowshoes, or by the elaborate etchings of an environmental artist – so we cannot infer something simply because it is a possible explanation. It must somehow be the best of competing explanations. These sorts of explanatory inferences are extremely common. The sleuth infers that the butler did it, since this is the best explanation of the evidence before him. The doctor infers that his patient has measles, since this is the best explanation of the symptoms. The astronomer infers the existence and motion of Neptune, since that is the best explanation of the observed perturbations of Uranus. Chomsky infers that our language faculty has a particular structure because this provides the best explanation of the way we learn to speak. Kuhn infers that normal science is governed by exemplars, since they provide the best explanation for the observed dynamics of research. This suggests a new model of induction, one that binds explanation and inference in an intimate and exciting way. According to Inference to the Best Explanation, our inferential practices are governed by explanatory considerations. Given our data and our background beliefs, we infer what would, if true, provide the best of the competing explanations we can generate of those data (so long as the best is good enough for us to make any inference at all).

Inference to the Best Explanation has become extremely popular in philosophical circles, discussed by many and endorsed without discussion by many more (For discussions, see, e.g., Pierce, 1931, 5.180-5.212, esp. 5.189; Harman, 1965; Brody, 1970; Hanson, 1972, ch. 4; Thagard, 1978; Cartwright, 1983, essay 5). Yet it still remains much more of a slogan than an articulated account of induction. In the balance of this section, I will make some first steps towards improving this situation. In the next section, we will consider the initial attractions of the view, as well as some apparent liabilities. The balance of this book is devoted to the questions of whether Inference to the Best Explanation really will provide an illuminating model of our inductive practices and whether it is an improvement over the other accounts we have considered.

The obvious way to flesh out Inference to the Best Explanation would be to insert one of the standard models of explanation. This, however, yields disappointing results, because of the backward state of those models. For example, we wouldn't get very far if we inserted the deductive-nomological model, since this would just collapse Inference to the Best Explanation into a version of the hypothetico-deductive model of confirmation. Indeed one suitable acid test for Inference to the Best Explanation is that it mark an improvement over the hypothetico-deductive model. As we saw in chapter one, the deductive-nomological model of explanation has many unattractive features; it also provides almost no resources for saying when one explanation is better than another. We will do better with the causal model of contrastive explanation I developed in the last chapter, as we will see in chapters five and six, but for now we are better off not burdening Inference to the Best Explanation with the details of any specific model of explanation, trying instead to stick to the actual explanatory relation itself, whatever its correct description turns out to be. Let us begin to flesh out the account by developing two signal distinctions that do not depend on the details of explanation: the distinction between actual and potential explanations, and the distinction between the explanation best supported by the evidence, and the explanation that would provide the most understanding or, in short, between the likeliest and the loveliest explanation.

Our discussion of inference, explanation, and the connection between the two is being conducted under the assumption of inferential and explanatory realism, an assumption we will not call into question until the final chapter of this book. I am assuming that a goal of inference is truth, that our actual inferential practices are truth-tropic, i.e. that they generally take us towards this goal, and that for something to be an actual explanation, it must be (at least approximately) true. But Inference to the Best Explanation cannot then be understood as inference to the best of the *actual* explanations. Such a model would make us too good at inference, since it would make all our inferences true. Our inductive practice is fallible: we sometimes reasonably infer falsehoods. This model would also fail to account for the role of competing explanations in inference. These competitors are typically incompatible and so cannot all be true, so we cannot represent them as competing actual explanations. The final and most important reason why Inference to the Best Actual Explanation could not describe our inductive practices is that it would not characterize the process of inference in a way we could follow, since we can only tell whether something is an actual explanation *after* we have settled the inferential question. It does not give us what we want, which is an account of the way explanatory considerations can serve as a guide to the truth. Telling someone to infer actual explanations is like a dessert recipe that says start with a soufflé. We are trying to describe the way we go from evidence to inference, but Inference to the Best Actual Explanation would require us already to have arrived in order to get there. In short, the model would not be epistemically effective.

The obvious solution, then, is to distinguish actual from *potential* explanation, and to construe Inference to the Best Explanation as Inference to the Best Potential Explanation. We have to produce a pool of potential explanations, from which we infer the best one. Although our discussion of explanation in the last two chapters considered only actual explanations, the distinction between actual and potential explanations is familiar in the literature on explanation. The standard version of the deductive-nomological model gives an account of potential explanation: there is no requirement that the explanation be true, only that it include a general hypothesis and entail the phenomenon. If we then add a truth requirement, we get an account of actual explanation (Hempel 1965, p. 338). By shaving the truth requirement off explanation, we seem to get a notion suitable for Inference to the Best Explanation: one that allows for the distinction between warranted and successful inferences, permits the competition between explanations to take place among incompatible hypotheses, and gives an account that is epistemically effective. According to Inference to the Best Explanation, then, we do not infer the best actual explanation; rather we infer that the best of the available potential explanations is an actual explanation.

The intuitive idea of a potential explanation is of something that satisfies all the conditions of an actual explanation, except possibly that of truth (Hempel, 1965, p. 338). This characterization may, however, be somewhat misleading, since it seems to entail the false generalization that all true potential explanations are actual explanations. The generalization does not hold for explanations that fit the deductive-nomological model, since a lawlike statement could hold in one possible world as a law, but in another as a mere coincidence. Even more clearly, it does not hold in the context of a causal model. A potential cause may exist yet not be an actual cause, say because some other cause pre-empted it. Of course one could construct a technical notion of potential explanation that satisfied the equality between true potential explanation and actual explanation, but this would not be a suitable notion for Inference to the Best Explanation. As the literature on Gettier cases shows, we often infer potential causes that exist but are not actual causes (Gilbert Harman's two-candle case is a good example of this; see Harman, 1973, pp. 22–3).

So we may need to do more work to characterize the notion of potential explanation that is suitable for Inference to the Best Explanation. One issue is how large we should make the pool. We might say that a potential explanation is any account that is logically compatible with all our observations (or almost all of them) and that is a possible explanation of the relevant phenomena. In other words, the potential explanations of some phenomena are those that do explain them in a possible world where our observations hold. This pool is very large, including all sorts of crazy explanations nobody would seriously consider. On the other hand, we might define the pool more narrowly, so that the potential explanations are only the 'live options': the serious candidates for an actual explanation. The advantage of the second characterization is that it seems to offer a better account of our actual procedure. When we decide which explanation to infer, we often start from a group of plausible candidates, and then consider which of these is the best, rather than selecting directly from the vast pool of possible explanations. In later chapters, we will see further reasons for preferring this restricted notion of potential explanation. But it is important to notice that the live options version of potential explanation already assumes an epistemic filter that limits the pool of potential explanations to plausible candidates. This version of Inference to the Best Explanation thus includes two filters, one that selects the plausible candidates, and a second that selects from among them. This view has considerable verisimilitude, but a strong version of Inference to the Best Explanation will not take the first filter as an unanalyzed mechanism, since epistemic filters are precisely the mechanisms that Inference to the Best Explanation is supposed to illuminate.

Let us turn now to the second distinction. It is important to distinguish two senses in which something may be the best of competing potential explanations. We may characterize it as the explanation that is most warranted: the 'likeliest' explanation. On the other hand, we may characterize the best explanation as the one which would, if correct, be the most explanatory or provide the most understanding: the 'loveliest' explanation. The criteria of likeliness and loveliness may well pick out the same explanation in a particular competition, but they are clearly different sorts of standard. Likeliness speaks of truth; loveliness of potential understanding. Moreover, the criteria do sometimes pick out different explanations. Sometimes the likeliest explanation is not very enlightening. It is extremely likely that smoking opium puts people to sleep because of its dormative powers (though not quite certain: it might be the oxygen that the smoker inhales with the opium, or even the depressing atmosphere of the opium den), but this is the very model of an unlovely explanation. An explanation can also be lovely without being likely. Perhaps some conspiracy theories provide examples of this. By showing that many apparently unrelated events flow from a single source and many apparent coincidences are really related, such a theory may have considerable explanatory power. If only it were true, it would provide a very good explanation. That is, it is lovely. At the same time, such an explanation may be very unlikely, accepted only by those whose ability to weigh evidence has been tilted by paranoia.

One of the reasons that likeliness and loveliness sometimes diverge is that likeliness is relative to the total available evidence, while loveliness is not, or at least not in the same way. We may have an explanation that is both lovely and likely given certain evidence, unlikely given additional evidence, yet still a lovely explanation of the original evidence. Newtonian mechanics is one of the loveliest explanations in science and, at one time, it was also very likely. More recently, with the advent of special relativity and the new data that support it, Newtonian mechanics has become less likely, but it remains as lovely an explanation of the old data as it ever was. Another reason for the divergence is that the two criteria are differently affected by additional competition. A new competitor may decrease the likeliness of an old hypothesis, but it will usually not change its loveliness. Even without the evidence that favored special relativity, the production of the theory probably made Newtonian mechanics less likely but probably not less lovely.

This gives us two more versions of Inference to the Best Explanation to consider: Inference to the Likeliest Potential Explanation and Inference to the Loveliest Potential Explanation. Which should we choose? There is a natural temptation to plump for likeliness. After all, Inference to the Best Explanation is supposed to describe strong inductive arguments, and a strong inductive argument is one where the premises make the conclusion likely. But in fact this connection is too close and, as a consequence, choosing likeliness would push Inference to the Best Explanation towards triviality. We want a model of inductive inference to describe what principles we use to judge one inference more likely than another, so to say that we infer the likeliest explanation is not helpful. To put the point another way, we want our account of inference to give the *symptoms* of likeliness, the features an argument has that lead us to say that the premises make the conclusion likely. A model of Inference to the

Likeliest Explanation begs these questions. It would still have some content, since it suggests that inference is a matter of selection from among competitors and that inference is often inference to a cause. But for Inference to the Best Explanation to provide an illuminating account, it must say more than that we infer the likeliest cause (cf. Cartwright, 1983, p. 6). This gives us a second useful acid test for Inference to the Best Explanation. (The first one was that it must do better than the hypothetico-deductive model.) Inference to the Best Explanation is an advance only if it reveals more about inference than that it is often inference to the likeliest cause. It should show how likeliness is determined (at least in part) by explanatory considerations.

So the version of Inference to the Best Explanation we should consider is Inference to the Loveliest Potential Explanation. Here at least we have an attempt to account for epistemic value in terms of explanatory virtue. This version claims that the explanation that would, if true, provide the deepest under-standing is the explanation that is likeliest to be true. Such an account suggests a really lovely explanation of our inferential practice itself, one that links the search for truth and the search for understanding in a fundamental way. Similar remarks apply to the notion of potential explanation, if we opt for the narrower live option characterization I favor. We want to give an account of the plausibility filter that determines the pool of potential explan-ations, and a deep version of Inference to the Best Explanation will give this characterization in explanatory terms: it will show how explanatory considerations determine plausibility.

The distinction between likeliness and loveliness is, I hope, reasonably clear. Nevertheless, it is easy to see why some philosophers may have conflated them. After all, if Inference to the Loveliest Explanation is a reasonable account, loveliness and likeliness will tend to go together, and indeed loveliness will be a guide to likeliness. Moreover, given the darkness of our 'inference box', we may be aware only of inferring what seems likeliest even if the mechanism actually works by assessing loveliness. Our awareness of what we are doing may not suggest the correct description. In any event, if there is a tendency to conflate the distinction, this helps to explain why Inference to the Best Explanation enjoys more popularity among philosophers than is justified by the arguments given to date in its favor. By im-plicitly construing the slogan simply as inference to the likeliest explanation, it is rightly felt to apply to a wide range of inferences; by failing to notice the difference between this and the deep account, the triviality is suppressed. At the same time, the distinction between likeliness and loveliness, or one like it, is one that most people who were seriously tempted to develop the account would make, and this may help to explain why the temptation has been so widely resisted. Once one realizes that an interesting version requires an account of explanatory loveliness that is conceptually independent of likeliness, the weakness of our grasp on what makes one explanation lovelier than another is discouraging.

In practice, these two versions of Inference to the Best Explanation are probably ideal cases: a defensible version may well need to combine elements of each, accounting for likeliness only partially in explanatory terms. For example, one might construct a version where a non-explanatory notion of likeliness plays a role in restricting membership in the initial set of potential explanations, but where considerations of loveliness govern the choice from among the members of that set. Again, we may have to say that considerations of likeliness having nothing to do with explanation will, under various conditions, defeat a preference for loveliness. This may be the only way to account for the full impact of disconfirming evidence. So the distinction between likeliness and loveliness leaves us with considerable flexibility. But I think we may take it as a rule of thumb that the more we must appeal to likeliness analyzed in non-explanatory terms to produce a defensible version of Inference to the Best Explanation, the less interesting that model is. Conversely, the more use we can make of the explanatory virtues, the closer we will come to fulfilling the exciting promise of Inference to the Best Explanation, of showing how explanatory considerations are our guide to the truth.

We have now gone some way towards spelling out the slogan, by making the distinctions between potential and actual explanation and between the likeliest and the loveliest explanation. By seeing how easy it is to slide from loveliness to likeliness, we have also sensitized ourselves to the risk of trivializing the model by making it so flexible that it can be used to describe almost any form of inference. But there are also various respects in which the scope of Inference to the Best Explanation is greater than may initially appear. Two apparent and damaging consequences of

Inference to the Best Explanation are that only one explanation can be inferred from any set of data and that the only data that are relevant to a hypothesis are data the hypothesis explains. Both of these are, however, merely apparent consequences, on a reasonable version of Inference to the Best Explanation. The first is easily disposed of. Inference to the Best Explanation does not require that we infer only one explanation of the data, but that we infer only one of *competing* explanations. The data from a flight recorder recovered from the wreckage of an airplane crash may at once warrant explanatory inferences about the motion of the plane, atmospheric conditions at the time of the accident, malfunctions of equipment in the airplane, and the performance of the pilot, and not simply because different bits of information from the recorder will warrant different inferences, but because the same bits may be explained in many different but compatible ways. When I notice that my front door has been forced open, I may infer both that I have been robbed and that my deadbolt is not as force-resistant as the locksmith claimed.

Inference to the Best Explanation can also account for some of the ways evidence may be relevant to a hypothesis that does not explain it. The most obvious mechanism for this depends on a deductive consequence condition on inference. If I am entitled to infer a theory, I am also entitled to infer whatever follows deductively from that theory, or from that theory along with other things I reasonably believe (cf. Hempel, 1965, pp. 31–2). This is at least highly plausible: it would be a poor joke to say one is entitled to believe a theory but not its consequences. Suppose now that I use Inference to the Best Explanation to infer from the data to a high-level theory, and then use the consequence condition to deduce a lower-level hypothesis from it. There is now no reason to suppose that the lower-level theory will explain all of the original data that indirectly support it. Newton was entitled to infer his theory in part because it explained the result of various terrestrial experiments. This theory in turn entails laws of planetary orbit. Inference to the Best Explanation with its consequence condition counts those laws as supported by the terrestrial evidence, even though the laws do not explain that evidence. It is enough that the higher-level theory does so. The clearest cases of the consequence condition, however, are deduced predictions. What will happen does not explain what happened in the past, but a theory that entails the prediction may.

Since Inference to the Best Explanation will sometimes underwrite inferences to high-level theories, rich in deductive consequences, the consequence condition substantially increases the scope of the model. Even so, we may wish to broaden the scope of the condition to include 'explanatory consequences' as well as strictly deductive ones. Seeing the distinctive flash of light, I infer that I will hear thunder. The thunder does not explain the flash, but the electrical discharge does and would also explain why I will hear thunder. But the electrical discharge does not itself entail that I will hear thunder. It is not merely that there is more to the story, but that there is always the possibility of interference. I might go temporarily deaf, the lightning may be too far away, I might sneeze just at the wrong moment, and there are always other possibilities that I do not know about. Someone who favors deductive models will try to handle these possibilities by including extra premises, but this will always require an unspecifiable *ceteris paribus* clause. So in many cases, it may be more natural to allow 'Inference from the Best Explanation' (Harman, 1986, pp. 68–70). Noticing that it is extraordinarily cold this morning, I infer that my car will not start. The failure of my car to start would not explain the weather, but my inference is naturally described by saying that I infer that it will not start because the weather would provide a good explanation of this, even though it does not entail it. (The risk of interference also helps to explain why we are often more confident of inferences to an explanation than inferences from an explanation. When we start from the effect, we know that there was no effective interference.)

ATTRACTIONS AND REPULSIONS

We have said enough to give some content to the idea of Inference to the Best Explanation. What the account now needs is some specific argument on its behalf, which I will begin to provide in the next chapter. First, however, it will be useful to compile a brief list of its general and *prima facie* advantages and disadvantages, some of which have already been mentioned, to prepare the ground for a more detailed assessment. Inference to the Best Explanation seems itself to be a relatively lovely explanation of our inductive practices. It gives a natural description of familiar aspects of our inferential procedures. The simplest reason for this is that we are often aware that we are inferring an explanation of the evidence, but there is more to it than that. We are also often aware of making an inferential choice between competing explanations, and this typically works by means of the two-filter process my favored version of Inference to the Best Explanation describes. We begin by considering plausible candidate explanations and then try to find data that discriminate between them. The account reflects the fact that a hypothesis that is a reasonable inference in one competitive milieu may not be in another. An inference may be defeated when someone suggests a better alternative explanation, even though the evidence does not change. Inference to the Best Explanation also suggests that we assess candidate inferences by asking a subjunctive question: we ask how good the explanation *would* be if it were true. There seems to be no reason why an inferential engine has to work in this way. If induction really did work by simple extrapolation, it would not involve subjunctive assessment. We can also imagine an inductive technique that included selection from among competitors but did not involve the subjunctive process. We might simply select on the basis of some feature of the hypotheses that we directly assess. In fact, however, we do often make the inductive decision whether something is true by asking what would be the case if it were, rather than simply deciding which is the likeliest possibility. We construct various causal scenarios and consider what they would explain and how well. Why is my refrigerator not running? Perhaps the fuse has blown. Suppose it has; but then the kitchen clock should not run either, since the clock and refrigerator are on the same circuit. Is the clock running? By supposing for the moment that a candidate explanation is correct, we can work out what further evidence is relevant to our inference. The role of subjunctive reasoning is partially captured by the familiar observation about the 'priority of theory over data'. Induction does not, in general, work by first gathering all the relevant data and only then considering the hypotheses to which they apply, since we often need to entertain a hypothesis first in order to determine what evidence is relevant to it (Hempel, 1966, pp. 12–13). But the point about subjunctive evaluation is not only that explanatory hypotheses are needed to determine evidential re-levance, but also a partial description of how that determination is made. (One of the attractions of the hypothetico-deductive model is that it also captures this subjunctive aspect of our methods for assessing relevant evidence, since we determine what a hypothesis entails by asking what would have to be the case if the hypothesis were true.)

Although we often infer an explanation just because that is where our interests lie, Inference to the Best Explanation correctly suggests that explanatory inferences should be common even in cases where explaining is not our primary purpose. Even when our main interest is in accurate prediction or effective control, it is a striking feature of our inferential practice that we often make an 'explanatory detour'. If I want to know whether my car will start tomorrow, my best bet is to try to figure out why it sometimes hasn't started in the past. When Semmelweis wanted to control the outbreak of childbed fever in one of the maternity wards in the Vienna hospital where he worked, he proceeded by trying to explain why the women in that ward were contracting the disease, and especially the contrast between the women in that ward and the women in another ward in the same hospital, who rarely contracted it. The method of explanatory detour seems to be one of the sources of the great predictive and manipulative successes of many areas of science. In science, the detour often requires 'vertical' inference to explanations in terms of unobserved and often unobservable entities and processes, and Inference to the Best Explanation seems particularly well equipped to account for this process.

In addition to giving a natural description of these various features of our inferential practice, Inference to the Best Explanation has a number of more abstract attractions. The notion of explanatory loveliness, upon which an interesting version of Inference to the Best Explanation relies, should help to make sense of the common observation of scientists that broadly aesthetic considerations of theoretical elegance, simplicity, and unification are a guide to inference. More generally, as I have already mentioned, the account describes a deep connection between our inferential and explanatory behavior, one that accounts for the prevalence of explanatory inferences even in cases where our main interests lie elsewhere. As such, it also helps with one of the problems of justifying our explanatory practices, since it suggests that one of the reasons for our obsessive search for explanations is that this is a peculiarly effective way of discovering the structure of the world. The explicit point of explaining is to understand *why* something is the case but, if Inference to the Best Explanation is correct, it is also our primary tool for discovering *what* is the case.

Another sort of advantage to the view that induction is Inference to the Best Explanation is that it avoids some of the objections to competing models of inductive inference or confirmation that we discussed in chapter one. One of the weaknesses of the simple Humean extrapolation model ('More of the Same') is that we are not always willing to extrapolate and, when we are, the account does not explain which of many possible extrapolations we actually choose. Inference to the Best Explanation does not always sanction extrapolation, since the best explanation for an observed pattern is not always that it is representative (Harman, 1965, pp. 90–1). Given my background knowledge, the hypothesis that I always win is not a good explanation of my early successes at the roulette wheel. Similarly, given a finite number of points on a graph marking the observed relations between two quantities, not every curve through those points is an equally good explanation of the data. One of the severe limitations of both the extrapolation view and the instantial model of confirmation is that they do not cover vertical inferences, where we infer from what we observe to something at a different level that is often unobservable. As we have seen, Inference to the Best Explanation does not have this limitation: it appears to give a univocal account of horizontal and vertical inferences, of inferences to what is observable and to the unobservable. The instantial model is also too permissive, since it generates the raven paradox. In chapter six, I will attempt to show how Inference to the Best Explanation helps to solve it.

Inference to the Best Explanation also seems a significant advance over the hypothetico-deductive model. First, while that model has very little to say about the 'context of discovery', the mechanisms by which we generate candidate hypotheses, the two-filter version of Inference to the Best Explanation suggests that explanatory considerations may apply to both the generation of candidates and the selection from among them. Second, since the deductive model is an account of confirmation rather than inference, it does not say when a hypothesis may actually be inferred. Inference to the Best Explanation does better here, since it brings in competition selection. Third, while the hypothetico-deductive model allows for vertical inference, it does not say much about how 'high' the inference may legitimately go. It allows the evidence to confirm a hypothesis however distant from it, so long as auxiliary premises can be found linking the two. In the next chapter, I will argue that Inference to the Best Explanation rightly focuses the impact of evidence more selectively, so that only some hypotheses that can be made to entail the evidence are supported by it. Fourth, if the model limits auxiliary hypotheses to independently known truths, it is too restrictive, since evidence may support a hypothesis even though the evidence is not entailed by the hypothesis and those auxiliaries, and it may disconfirm a hypothesis without contradicting it. Inference to the Best Explanation allows for this sort of evidence, since explanation does not require deduction. Finally, Inference to the Best Explanation avoids several of the sources of over-permissiveness that are endemic to the deductive model. In addition to avoiding the raven paradox, Inference to the Best Explanation blocks confirmation by various non-explanatory deductions. One example is the confirmation of an arbitrary conjunction by a conjunct, since a conjunction does not explain its conjuncts. For these as well as for some of the other liabilities of the hypothetico-deductive model, a symptom of the relative advantages of Inference to the Best Explanation is that many of the problems of the hypothetico-deductive model of confirmation and of the deductive-nomological model of explanation 'cancel out': many of the counterexamples of the one are also counterexamples of the other. This suggests that the actual explanatory relation offers an improved guide to inference.

We have so far canvassed two sorts of advantages of Inference to the Best Explanation. The first is that it is itself a lovely explanation of various aspects of inference; the second is that it is better than the competition. The third and final sort of advantage I will mention is that, in addition to accounting for scientific and everyday inference, Inference to the Best Explanation has a number of distinctively philosophical applications. The first is that it accounts for its own discovery. In chapter one, I suggested that the task of describing our inductive behavior is itself a broadly inductive project, one of going from what we observe about our inferential practice to the mechanism that governs it. If this is right, a model of induction ought to apply to itself. Clearly the extrapolation and the instantial models do not do well on this criterion, since the inference is to the contents of a black box, the sort of vertical inference those models do not sanction. Nor does the hypothetico-deductive model do much better, since it does not entail many observed features of our practice. It does not, for example, entail any of the inferences we actually make. Inference to the Best Explanation does much better on this score. The inference to an account of induction is an explanatory inference: we want to explain why we make the inferences we do. Our procedure has been to begin with a pool of plausible candidate explanations (the various models of induction we have canvassed) and then select the best. This is a process of competition selection which works in part by asking the subjunctive question of what sort of inferences we would make, if we used the various models. Moreover, if we do end up selecting Inference to the Best Explanation, it will not simply be because it seems the likeliest explanation, but because it has the features of unification, elegance, and simplicity that make it the loveliest explanation of our inductive behavior.

Another philosophical application of Inference to the Best Explanation is to the local justification of some of our inferential practices. For example, it is widely supposed that a theory is more strongly supported by successful predictions than by data that were known before the theory was constructed and which the theory was designed to accommodate. At the same time, the putative advantage of prediction over accommodation is controversial and puzzling, because the logical relationships between theory and data upon which inductive support is supposed to depend seem unaffected by the merely historical fact of when the data were observed. But there is a natural philosophical inference to the best explanation that seems to defend the epistemic distinction. When data are predicted, the best explanation for the fit between theory and data, it is claimed, is that the theory is true. When the data are accommodated, however, there is an alternative explanation of the fit, namely that the theory was designed just for that purpose. This explanation, which only applies in the case of accommodation, is better than the truth explanation, and so Inference to the Best Explanation shows why prediction is better than accommodation (cf. Horwich, 1982, p. 111). We will assess this argument in chapter eight.

Another example of the application of Inference to the Best Explanation to local philosophical justification is in connection with Thomas Kuhn's notorious discussion of 'incommensurability'

71

(Kuhn, 1970, esp. chs 9–10). According to him, there is no straightforward way of resolving scientific debates during times of 'scientific revolutions', because the disputants disagree about almost everything, including the evidence. This seems to block resolution by any appeal to a crucial experiment. On the traditional view of such experiments, they resolve theoretical disputes by providing evidence that simultaneously refutes one theory while supporting the other. Competing theories are found to make conflicting predictions about the outcome of some experiment; the experiment is performed and the winner is determined. But this account seems to depend on agreement about the outcomes of experiment, which Kuhn denies. These experiments can, however, be redescribed in terms of Inference to the Best Explanation in a way that does not assume shared observations. A crucial experiment now becomes two experiments, one for each theory. The outcome of the first experiment is explained by its theory, whereas the outcome of the second is not explained by the other theory, so we have some basis for a preference. Shared standards of explanation may thus compensate for observational disagreement: scientists should prefer the theory that best explains its proper data. There is, however, more to Kuhn's notion of incommensurability than disagreement over the data; in particular, there is also tacit disagreement over explanatory standards. But this may turn out to be another advantage of Inference to the Best Explanation. Insofar as Kuhn is right here, Inference to the Best Explanation will capture the resulting indeterminacy of scientific debate that is an actual feature of our inferential practices.

Another well-known philosophical application of Inference to the Best Explanation is to argue for various forms of realism. For example, as part of an answer to the Cartesian skeptic who asks how we can know that the world is not just a dream or that we are not just brains in vats, the realist may argue that we are entitled to believe in the external world since hypotheses that presuppose it provide the best explanation of our experiences. It is possible that it is all a dream, or that we are really brains in vats, but these are less good explanations of the course of our experiences than the ones we all believe, so we are rationally entitled to our belief in the external world. There is also a popular application of Inference to the Best Explanation to realism in the philosophy of science, which we have already briefly mentioned. The issue here is whether scientific theories, particularly those that appeal to unobservables, are getting at the truth, whether they are providing an increasingly accurate representation of the world and its contents. There is an inference to the best explanation for this conclusion. In brief, later theories tend to have greater predictive success than those they replace, and the best explanation for this is that later theories are better descriptions of the world than earlier ones. We ought to infer scientific realism, because it is the best explanation of predictive progress. We will assess this argument in chapter nine.

Let us conclude this chapter with some of the bad news. First, several of the philosophical applications of Inference to the Best Explanation can be questioned. In the case of the argument for the advantages of prediction over accommodation, one may ask whether the 'accommodation explanation' really competes with the truth explanation (Horwich, 1982, pp. 112–16). If not, then as we saw in the last section, Inference to the Best Explanation does not require that we choose between them. Moreover, the assumption that they do compete, that explaining the fit between theory and accommodated data by appeal to the act of accommodation pre-empts explaining the fit by appeal to the truth of the theory, seems just to assume that accommodation does not provide support, and so to beg the question. As for the argument for realism about the external world, do our beliefs about the world really provide a better explanation than the dream hypothesis, or is it simply that this is the explanation we happen to prefer? Again, doesn't the inference to scientific realism as the best explanation for predictive success simply assume that inferences to the best explanation are guides to the truth about unobservables, which is just what an opponent of scientific realism would deny? A second sort of liability is the suspicion that Inference to the Best Explanation is still nothing more than Inference to the Likeliest Cause in fancy dress, and so fails to account for the symptoms of likeliness. Third, insofar as there is a concept of explanatory loveliness that is conceptually distinct from likeliness, one may question whether this is a suitable criterion of inference. On the one hand, there is what we may call 'Hungerford's objection', in honor of the author of the line, 'Beauty is in the eye of the beholder'. Perhaps explanatory loveliness is too subjective and interest relative to give an account of inference that reflects the objective features of inductive warrant.

On the other hand, supposing that loveliness is as objective as inference, we have 'Voltaire's objection'. What reason is there to believe that the explanation that would be loveliest, if it were true, is also the explanation that is most likely to be true? Why should we believe that we inhabit the loveliest of all possible worlds? As we saw in the last section, Inference to the Best Explanation requires that we work with a notion of potential explanation that does not carry a truth requirement. Once we have removed truth from explanation, however, it is not clear how we get it back again (cf. Cartwright, 1983, pp. 89–91). Lastly, perhaps most importantly, it will be claimed that Inference to the Best Explanation is only as good as our account of explanatory loveliness, and this account is nonexistent. In the next chapter, I begin to meet this objection.

5

CONTRASTIVE INFERENCE

A CASE STUDY

In this chapter and the next two, I will consider some of the prospects of Inference to the Best Explanation as a solution to the descriptive problem of inductive inference. We want to determine how illuminating that account is as a description, or a partial description, of the mechanism inside the cognitive black box that governs our inductive practices. To do this, we need to show how explanatory considerations are a guide to inference, how loveliness helps to determine likeliness. In particular, we want to see whether the model can meet the two central challenges from the last chapter, to show that inferences to the best explanation are more than inferences to the likeliest cause, and to show that Inference to the Best Explanation marks an advance over the simple hypothetico-deductive model.

As I have stressed, the main difficulty standing in the way of this project is our poor understanding of what makes one explanation lovelier than another. Little has been written on this subject, perhaps because it has proven so difficult even to say what makes something an explanation. How can we hope to determine what makes one explanation better than another, if we can't even agree about what distinguishes explanations of any quality from things that are not explanations at all? Moreover, most of what has been written about explanatory loveliness has focused on the interest relativity of explanation, which seems to bring out pragmatic and subjective factors that are too variable to provide a suitably objective measure of inductive warrant.

Yet the situation is not hopeless. My analysis of contrastive explanation in chapter three will help. There I argued that pheno-

75

mena we explain often have a contrastive fact–foil structure, and that the foil helps to select the part of the causal history of the fact that provides a good explanation by means of a mechanism of causal triangulation. According to my Difference Condition, to explain why P rather than Q, we need to find a causal difference between P and not-Q, consisting of a cause of P and the absence of a corresponding event in the history of not-Q. Thus we can explain why Jones rather than Smith contracted paresis by pointing out that only Jones had syphilis, since this is to point out a causal difference between the two men, even though most people with syphilis do not get paresis. This account of contrastive explanation shows how what counts as a good explanation depends on interests, since interests determine the choice of foil, and a cause that marks a difference for one foil will not generally do so for another. Jones's syphilis would not explain why he rather than Doe (who also had syphilis) contracted paresis; here the explanation might be instead that only Jones left his syphilis untreated. In this respect, then, what counts as a lovely explanation of P depends on one's interests, but this cashes out into the question of whether the cited cause provides any explanation at all of the contrast that expresses a particular interest.

The sensitivity of explanation to choice of foils captures much of what has been said about interest relativity, and it also shows that these factors are not strongly subjective in a way that would make them irrelevant to inference. An account of the interest relativity of explanation would be strongly subjective if it showed that what counts as a good explanation depends on the tastes of the audience rather than the causal structure of the world. Convincing examples of this would have to be cases where different people favor incompatible explanations of the same phenomenon. It is no threat to the objectivity of explanation that different people should be interested in explaining different phenomena, and it is obvious that a good explanation of one phenomenon is not usually a good explanation of another. A contrastive analysis of explanation only supports this innocuous form of relativity, if we construe the phenomena themselves as contrastive, so that a change in foil yields a different phenomenon. Moreover, my analysis of contrastive explanation shows that a change in foil helps to select a different part of the same causal history. Differences in interest require different but compatible explanations, and this does not bring in strong subjectivity. And this much interest relativity is also something any reasonable account of inference must acknowledge: different people may all reasonably infer different things from shared evidence, depending on their inferential interests, when the inferences are compatible.

So my account of contrastive explanation helps to defuse the objection to Inference to the Loveliest Explanation, that loveliness is hopelessly subjective. (We will return to this issue in chapter seven.) It also provides the core of a positive account of one way explanatory considerations can serve as a guide to inference. The reason for this is the structural similarity between the Difference Condition and Mill's Method of Difference. According to Mill, we find the cause of a fact in some prior difference between a case where the fact occurs and an otherwise similar case where it does not. Mill's central mechanism for inferring the likeliest cause is almost the same as the mechanism of causal triangulation which helps to determine the loveliest explanation. This near-isomorphism provides an important argument in favor of Inference to the Best Explanation, since it shows that a criterion we use to evaluate the quality of potential explanations is the same as one that is designed to maximize the chances that we correctly infer a cause. By inferring something that would provide a good explanation of the contrast if it were a cause, we are led to infer something that is likely to be a cause. Returning to poor Jones, we may find that his condition, taken alone, points to no particular explanation. But if we try instead to explain why Jones rather than Smith contracted paresis, we will be led, by means of the Difference Condition, to look for some possibly relevant difference in the medical histories of the two men. Thus we may infer that Jones's syphilis was a cause of his paresis, since this is an explanatory difference. And this is just where Mill's method would take us, if syphilis was the only possibly relevant difference. Moreover, our explanation and our inference will both change if we change the foil. If we ask why Jones rather than Doe contracted paresis, we will be led to explain this contrast by appealing to Doe's treatment. By varying the foil, we change the best explanation, and this leads us to different but compatible inductive inferences, taking us to different stages of Jones's medical history.

By considering inferences to contrastive explanations, we go some way towards meeting the challenge that Inference to the Best Explanation is nothing more than Inference to the Likeliest

Cause. Because looking for residual differences in similar histories of fact and foil is a good way of determining a likely cause, as Mill taught us, and because contrastive explanation depends on just such differences, looking for potential contrastive explanations can be a guide to causal inference. Given contrastive data, the search for explanation is an effective way of determining just what sort of causal hypotheses the evidence supports. This procedure focuses our inferences, by eliminating putative causes that are in the shared part of antecedents of fact and foil. These antecedents may well be causally relevant, but the fact that they would not explain the contrast shows that the contrast does not (at least by itself) provide evidence that they are causes. This version of Inference to the Best Explanation thus sheds some light on the context of discovery, since the requirement that a potential explanation cite a difference severely restricts the class of candidate hypotheses. It also brings out one role of background knowledge in inference in a natural way, since our judgment of which antecedents are shared, a judgment essential to the application of the method, will depend on such knowledge.

I also want to argue, in this chapter and the next, that Inference to the Best Contrastive Explanation helps to meet the second challenge, to show that the model is better than simple hypothetico-deductivism. It marks an improvement both where the deductive model is too strict, neglecting evidential relevance in cases where there is no appropriate deductive connection between hypothesis and data, and where it is too lenient, registering support where there is none to be had as, for example, the raven paradox shows. Inference to the Best Explanation does better in the first case because, as the analysis of contrastive explanation shows, explanatory causes need not be sufficient for their effects, so the fact that a hypothesis would explain a contrast may provide some reason to believe the hypothesis, even though the hypothesis does not entail the data. It does better in the second case because, while some contrapositive instances (e.g. non-black non-ravens) do support a hypothesis, not all do, and the requirement of shared antecedents helps to determine which do and which do not. The structural similarity between the Method of Difference and contrastive explanation that I will exploit in these chapters will also eventually raise the question of why Inference to the Best Explanation is an improvement on Mill's methods, a question I will address in chapter seven.

To develop these arguments and, more generally, to show just how inferences to contrastive explanations work, it is useful to consider a simple but actual scientific example in some detail. The example I have chosen is Ignaz Semmelweis's research from 1844 to 1848 on childbed fever, taken from Hempel's well-known and characteristically clear discussion (Hempel, 1966, pp. 3–8). Semmelweis wanted to find the cause of this often fatal disease, which was contracted by many of the women who gave birth in the Viennese hospital in which he did his research. His central datum was that a much higher percentage of the women in the First Maternity Division of the hospital contracted the disease than in the adjacent Second Division, and he sought to explain this difference. The hypotheses he considered fell into three types. In the first were hypotheses that did not mark differences between the divisions, and so were rejected. Thus, the theory of 'epidemic influences' descending over entire districts did not explain why more women should die in one division than another; nor did it explain why the mortality among Viennese women who gave birth at home or on the way to the hospital was lower than in the First Division. Similarly, the hypotheses that the fever was caused by overcrowding, by diet, or by general care were rejected because these factors did not mark a difference between the divisions.

One striking difference between the two divisions was that medical students only used the First Division for their obstetrical training, while midwives received their training in the Second Division. This suggested the hypothesis that the high rate of fever in the First Division was caused by injuries due to rough examination by the medical students. Semmelweis rejected the rough examination hypothesis on the grounds that midwives performed their examinations in more or less the same way, and that the injuries due to childbirth were in any case greater than those due to rough examination.

The second type of hypotheses included those that did mark a difference between the divisions, but where eliminating the difference in putative cause did not affect the difference in mortality. A priest delivering the last sacrament to a dying woman had to pass through the First Division to get to the sickroom where dying women were kept, but not through the Second Division. This suggested that the psychological influence of seeing the priest might explain the difference, but Semmelweis ruled this out by arranging for the priest not to be seen by the women in the First Division either and finding that this did not affect the mortality rates. Again, women in the First Division were delivered lying on their backs, while women in the Second delivered on their sides but, when Semmelweis arranged for all women to deliver on their sides, the mortality remained the same.

The last type of hypothesis that Semmelweis considered is one that marked a difference between the divisions, and where eliminating this difference also eliminated the difference in mortality. Kolletschka, one of Semmelweis's colleagues, received a puncture wound in his finger during an autopsy, and died from an illness with symptoms like those of childbed fever. This led Semmelweis to infer that Kolletschka's death was due to the 'cadaveric matter' that the wound introduced into his bloodstream, and Semmelweis then hypothesized that the same explanation might account for the deaths in the First Division, since medical students performed their examinations directly after performing autopsies, and midwives did not perform autopsies at all. Similarly, the cadaveric hypothesis would explain why women who delivered outside the hospital had a lower mortality from childbed fever, since they were not examined. Semmelweis had the medical students disinfect their hands before examination, and the mortality rate in the First Division went down to the same low level as that in the Second Division. Here at last was a difference that made a difference, and Semmelweis inferred the cadaveric hypothesis.

Hempel's case study is a gold mine for inferences to the best contrastive explanation. Let us begin by considering Semmelweis's strategy for each of the three groups of hypotheses, those of no difference, of irrelevant differences, and of relevant differences. His rejection of the hypotheses in the first group – epidemic influences, overcrowding, general care, diet, and rough examination – show how Inference to the Best Explanation can account for negative evidence. These hypotheses are rejected on the grounds that, though they are compatible with the evidence, they would not explain the contrast between the divisions. Epidemic influences, for example, still might possibly be part of the causal history of the deaths in the First Division, say because the presence of these influences is a necessary condition for any case of childbed fever. And nobody who endorsed the epidemic hypothesis would have claimed that the influences were sufficient for the fever, since it was common knowledge that not all mothers in the district contracted childbed fever. Still, Semmelweis took the fact that the hypotheses in the first group would not explain the contrast between the divisions or the contrast between the First Division and mothers who gave birth outside the hospital to be evidence against them.

Semmelweis also used a complementary technique for discrediting the explanations in the first group that is naturally described in terms of Inference to the Best Explanation, when he argued against the epidemic hypothesis on the grounds that the mortality rate for births outside the hospital was lower than in the First Division. What he had done was to change the foil, and point out that the hypothesis also fails to explain this new contrast. It explains neither why mothers get fever in the First Division rather than in the Second, nor why mothers get fever in the First Division rather than outside the hospital. Similarly, when Semmelweis argued against the rough examination hypothesis on the grounds that childbirth is rougher on the mother than any examination, he pointed out not only that it fails to explain why there is fever in the First Division rather than in the Second, but also why there is fever in the First Division rather than among other mothers generally. New foils provide new evidence, in these cases additional evidence against the putative explanations.

The mere fact that the hypotheses in the first group did not explain some evidence can not, however, account for Semmelweis's negative judgment. No hypothesis explains every observation, and most evidence that is not explained by a hypothesis is simply irrelevant to it. But Semmelweis's observation that the hypotheses do not explain the contrast in mortality between the divisions seems to count against those hypotheses in a way that, say, the observation that those hypotheses would not explain why the women in the First Division were wealthier than those in the Second Division (if they were) would not. Of course, since Semmelweis was interested in reducing the incidence of childbed fever, he was naturally more interested in an explanation of the contrast in mortality than in an explanation of the contrast in wealth, but this does not show why the failure of the hypotheses to explain the first contrast counts against them. This poses a general puzzle for Inference to the Best Explanation: how can that account distinguish negative evidence from irrelevant evidence, when the evidence is logically consistent with the hypothesis?

One straightforward mechanism is rival support. In some cases, evidence counts against one hypothesis by improving the explanatory power of a competitor. The fact that the mortality in the First Division went down when the medical students disinfected their hands before examination supports the cadaveric matter hypothesis, and so indirectly counts against all the hypotheses inconsistent with it that cannot explain this contrast. But this mechanism of disconfirming an explanation by supporting a rival does not seem to account for Semmelweis's rejection of the hypotheses in the first group, since at that stage of his inquiry he had not yet produced an alternative account.

Part of the answer to this puzzle about the difference in the epistemic relevance of a contrast in mortality and a contrast in wealth is that the rejected hypotheses would have enjoyed some support from the fact of mortality but not from the fact of wealth. The epidemic hypothesis, for example, was not Semmelweis's invention, but a popular explanation at the time of his research. Its acceptance presumably depended on the fact that it seemed to provide an explanation, if a weak one, for the non-contrastive observations of the occurrence of childbed fever. In the absence of a stronger and competing explanation, this support might have seemed good enough to justify the inference. But by pointing out that the hypothesis does not explain the contrast between the divisions, Semmelweis undermines this support. On the other hand, the epidemic hypothesis never explained, and so was never supported by observations about, the wealth of the victims of childbed fever, so its failure to explain why the women in the First Division were wealthier than those in the Second Division would not take away any support it had hitherto enjoyed.

On this view, the observation that the hypotheses in the first group do not explain the contrast in mortality and the observation that they do not explain the contrast in wealth are alike in that they both show that these data do not support the hypothesis. The difference in impact only appears when we take into account that only evidence about mortality had been supposed to support the hypothesis, so only in this case is there a net loss of support. This view seems to me to be correct as far as it goes, but it leaves a difficult question. Why, exactly, does the failure to explain the contrast in mortality undermine prior support for hypotheses in the first group? Those hypotheses would still give some sort of explanation for the cases of the fever in the hospital, even if they would not explain the contrast between the divisions. Consider a different example. Suppose that we had two wards of patients who suffered from syphilis and discovered that many more of them in one ward contracted paresis than in the other. The hypothesis that syphilis is a necessary cause of paresis would not explain this contrast, but this would not, I think, lead us to abandon the hypothesis on the grounds that its support had been undermined. Instead, we would continue to accept it and look for some further and complementary explanation for the difference between the wards, say in terms of a difference in the treatments provided. Why, then, is Semmelweis's case any different?

The difference must lie in the relative weakness of the initial evidence in support of the hypotheses in the first group. If the only evidence in favor of the epidemic hypothesis is the presence of childbed fever, the contrast in mortality does undermine the hypothesis, because it suggests that the correct explanation of the contrast will show that epidemic influences have nothing to do with fever. If, on the other hand, the epidemic hypothesis would also explain why there were outbreaks of fever at some times rather than others, or in some hospitals rather than others, even though these cases seemed similar in all other plausibly relevant respects, then we would be inclined to hold on to that hypothesis and look for a complementary explanation of the contrast between the divisions. In the case of syphilis and paresis, we presumably have extensive evidence that there are no known cases of paresis not preceded by syphilis. The syphilis hypothesis not only would explain why those with paresis have it, but also the many contrasts between people with paresis and those without it. This leads us to say that the correct explanation of the contrast between the wards is more likely to complement the syphilis hypothesis than to replace it.

If this account is along the right lines, then the strength of the disconfirmation provided by the failure to explain a contrast depends on how likely it seems that the correct explanation of the contrast will pre-empt the original hypothesis. This explains our different reaction to the wealth case. We may have no idea why the women in the First Division are wealthier than those in the Second, but it seems most unlikely that the reason for this will pre-empt the hypotheses of the first group. When we judge that pre-emption is likely, we are in effect betting that the best explanation of the contrast will either contradict the original hypothesis or show it to be unnecessary, and so that the evidence that originally supported it will instead support a competitor. So the mechanism here turns out to be an attenuated version of disconfirmation by rival support after all. The inability of the hypotheses in the first group to explain the contrast between the divisions and the contrast between the First Division and births outside the hospital disconfirms those hypotheses because, although the contrastive data do not yet support a competing explanation, since none has yet been formulated, Semmelweis judged that the best explanation of those contrasts would turn out to be a competing rather than a complementary account. This judgment can itself be construed as an overarching inference to the best explanation. If we reject the hypotheses in the first group because they fail to explain the contrasts, this is because we regard the conjecture that the hypotheses are wrong to be a better explanation of the failures than that they are merely incomplete. Judgments of this sort are speculative, and we may in the end find ourselves inferring an explanation of the contrasts that is compatible with the hypotheses in the first group, but insofar as we do take their explanatory failures to count against them, I think it must be because we do make these judgments.

On this view, given a hypothesis about the etiology of a fact, and faced with the failure of that hypothesis to explain a contrast between that fact and a similar foil, the scientist must choose between the overarching explanations that the failure is due to incompleteness and that it is due to incorrectness. Semmelweis's rejections of the hypotheses in the first group are examples of choosing the incorrectness explanation. It is further corroboration of the claim that these choices must be made that we cannot make sense of Semmelweis's research without supposing that he also sometimes inferred incompleteness. For while the cadaveric hypothesis had conspicuous success in explaining the contrast between the divisions, it failed to explain other contrasts that formed part of Semmelweis's evidence. For example, it did not explain why some women in the Second Division contracted childbed fever while others in that division did not, since none of the midwives who performed the deliveries in that division performed autopsies. Similarly, the cadaveric hypothesis did not explain why some women who had street births on their way to the hospital contracted the fever, since those women were rarely examined by either medics or midwives after they arrived. Consequently, if we take it that Semmelweis nevertheless had good reason to believe that infection by cadaveric matter was a cause of childbed fever, it can only be because he reasonably inferred that the best explanation of these explanatory failures was that the cadaveric hypothesis was incomplete, not the only cause of the fever, rather than that it was incorrect. These cases also show that we cannot in general avoid the speculative judgment by waiting until we actually produce an explanation for all the known relevant contrasts, since in many cases this would postpone inference indefinitely.

Let us turn now to the two hypotheses of the second group, concerning the priest and delivery position. Unlike the hypotheses of the first group, these did mark differences between the divisions and so might explain the contrast in mortality. The priest bearing the last sacrament only passed through the First Division, and only in that division did mothers deliver on their backs. Since these factors were under Semmelweis's control, he tested these hypotheses in the obvious way, by seeing whether the contrast in mortality between the divisions remained when these differences were eliminated. Since that contrast remained, even when the priest was removed from the scene and when the mothers in both divisions were delivered on their sides, these hypotheses could no longer be held to explain the original contrast.

This technique of testing a putative cause by seeing whether the effect remains when it is removed is widely employed. Semmelweis could have used it even without the contrast between the divisions, and it is worth seeing how a contrastive analysis could account for this. Suppose that all the mothers in the hospital were delivered on their backs, and Semmelweis tested the hypothesis that this delivery position is a cause of childbed fever by switching positions. He might have done this for only some of the women, using the remainder as a control. In this case, the two groups would have provided a potential contrast. If a smaller percentage of the women who were delivered on their sides contracted childbed fever, the delivery hypothesis would have explained and so been supported by this contrast. And even if Semmelweis had switched all the mothers, he would have had a potential diachronic (before and after) contrast, by comparing the incidence of fever before and after the switch. In either case, a contrast would have supported the explanatory inference. In fact, however, these procedures would not have produced a contrast, since delivery position is irrelevant to childbed fever. This absence of contrast would not disprove the delivery hypothesis. Delivering on the back might still be a cause of fever, though there might be some obscure alternate cause that came into play when the delivery position was switched. But the absence of the contrast certainly would disconfirm the delivery hypothesis. The reason for this is the same as in the case of the epidemic hypothesis: the likeliness of a better, pre-emptive explanation. Even if Semmelweis did not have an alternative explanation for the cases of fever, there must be another explanation in the cases of side delivery, and it is likely that this explanation will show that back delivery is irrelevant even when it occurs. As in the case of the hypotheses in the first group, when we take an explanatory failure to count against a hypothesis, even when we do not have an alternative explanation, this is because we infer that the falsity of the hypothesis is a better explanation for its explanatory failure than its incompleteness.

This leaves us with Semmelweis's final hypothesis, that the difference in mortality is explained by the cadaveric matter that the medical students introduced into the First Division. Here too we have an overarching explanation in play. Semmelweis had already conjectured that the difference in mortality was somehow explained by the fact that mothers were attended by medical students in the First Division, and by midwives in the Second Division. This had initially suggested the hypothesis that the rough examinations given by the medical students was the cause, but this neither explained the contrast between the divisions nor the contrast between the mothers in the First Division and mothers generally, who suffered more from labor and childbirth than from any examination. The cadaveric hypothesis was another attempt to explain the difference between the divisions under the overarching hypothesis that the contrast is due to the difference between medical students and midwives. In addition to explaining the difference between divisions, this hypothesis would explain Kolletschka's illness, as well as the difference between the First Division and births outside the hospital.

Finally, Semmelweis tested this explanation by eliminating the

cadaveric matter with disinfectant and finding that this eliminated the difference in the mortality between the divisions. This
too can be seen as the inference to a contrastive explanation for a
new contrast, where now the difference that is explained is not
the simple difference in mortality between the divisions, but the
diachronic contrast between the initial presence of that difference
and its subsequent absence. The best explanation of the fact that
removing the cadaveric matter was followed by the elimination
of the difference in mortality was that the cadaveric matter was
responsible for that difference. By construing Semmelweis's
evidence as a diachronic contrast, we bring out the important
point that the comparative data have a special probative force
that we would miss if we simply treated them as two separate
confirmations of Semmelweis's hypothesis.

Semmelweis's research into the causes of childbed fever brings
out many of the virtues of Inference to the Best Explanation when
that account is tied to a model of contrastive explanation. In
particular, it shows how explanatory considerations focus and
direct inquiry. Semmelweis's work shows how the strategy of
considering potential contrastive explanations focuses inquiry,
even when the ultimate goal is not simply an explanation.
Semmelweis's primary interest was to eliminate or at least reduce
the cases of childbed fever, but he nevertheless posed an
explanatory question: Why do women contract childbed fever?
His initial ignorance was such, however, that simply asking why
those with the fever have it did not generate a useful set of
hypotheses. So he focused his inquiry by asking contrastive
why-questions. His choice of the Second Division as foil was
natural because it provided a case where the effect is absent yet
the causal histories are very similar. By asking why the contrast
obtains, Semmelweis focused his search for explanatory
hypotheses on the remaining differences. This strategy is widely
applicable. If we want to find out why some phenomenon occurs,
the class of possible causes is often too big for the process of
Inference to the Best Explanation to get a handle on. If, however,
we are lucky or clever enough to find or produce a contrast where
fact and foil have similar histories, most potential explanations
are immediately 'cancelled out' and we have a manageable and
directed research program. The contrast will be particularly
useful if, as in Semmelweis's case, in addition to meeting the
requirement of shared history, it is also a contrast that various

available hypotheses will not explain. Usually, this will still leave more than one hypothesis in the field, but then further observation and experiment may produce new contrasts that leave only one explanation. This shows how the interest relativity of explanation is at the service of inference. By tailoring his explanatory interests (and his observational and experimental procedures) to contrasts that would help to discriminate between competing hypotheses, Semmelweis was able to judge which hypothesis would provide the best overall explanation of the wide variety of contrasts (and absences of contrast) he observed, and so to judge which hypothesis he ought to infer. Semmelweis's inferential interests determined his explanatory interests, and the best explanation then determined his inference.

EXPLANATION AND DEDUCTION

Semmelweis's research is a simple and striking illustration of inferences to the best explanation in action, and of the way they exploit contrastive data. It is also Hempel's paradigm of the hypothetico-deductive method. So this case is particularly well suited for a comparison of the virtues of Inference to the Best Explanation and the deductive model. The example, I suggest, shows that Inference to the Best Explanation is better than hypothetico-deductivism.

Consider first the context of discovery. Semmelweis's use of contrasts and prior differences to help generate a list of candidate hypotheses illustrates one of the ways Inference to the Best Explanation elucidates the context of discovery, a central feature of our inductive practice neglected by the hypothetico-deductive model. The main reason for this neglect is easy to see. Hypothetico-deductivists emphasize the hopelessness of 'narrow inductivism', the view that scientists ought to proceed by first gathering all the relevant data without theoretical preconception and then using some inductive algorithm to infer from those data to the hypothesis they best support. Scientists never have all the relevant data, they often cannot tell whether or not a datum is relevant without theoretical guidance, and there is no general algorithm that could take them from data to a hypothesis that refers to entities and processes not mentioned in the data (Hempel, 1966, pp. 10–18). The hypothetico-deductive alternative is that, while scientists never have all the data, they can at least determine relevance if the hypothesis comes first. Given a conjectural hypothesis, they know to look for data that either can be deduced from it or would contradict it. The cost of this account is that we are left in the dark about the source of the hypotheses themselves. According to Hempel, scientists need to be familiar with the current state of research, and the hypotheses they generate should be consistent with the available evidence but, in the end, generating good hypotheses is a matter of 'happy guesses' (Hempel, 1966, p. 15).

The hypothetico-deductivist must be right in claiming that there are no universally shared mechanical rules that generate a unique hypothesis from any given pool of data since, among other things, different scientists generate different hypotheses, even when they are working with the same data. Nevertheless, this 'narrow hypothetico-deductivist' conception of inquiry badly distorts the process of scientific invention. Most hypotheses consistent with the data are non-starters, and the use of contrastive evidence and explanatory inference is one way the field is narrowed. In advance of an explanation for some effect, we know to look for a foil with a similar history. If we find one, this sharply constrains the class of hypotheses that are worth testing. A reasonable conjecture must provide a potential explanation of the contrast, and most hypotheses that are consistent with the data will not provide this. (For hypotheses that traffic in unobservables, the restriction to potential contrastive explanations still leaves a lot of play: we will consider further ways the class of candidate hypotheses is restricted in chapter seven.)

The slogan 'Inference to the Best Explanation' may itself bring to mind an excessively passive picture of scientific inquiry, suggesting perhaps that we simply infer whatever seems the best explanation of the data we happen to have. But the Semmelweis example shows that the account, properly construed, allows for the feedback between the processes of hypothesis formation and data acquisition that characterizes actual inquiry. Contrastive data suggest explanatory hypotheses, and these hypotheses in turn suggest manipulations and controlled experiments that may reveal new contrasts that help to determine which of the candidates is the best explanation. This is one of the reasons the subjunctive element in Inference to the Best Explanation is important. By considering what sort of explanation the hypothesis would provide, if it were true, we assess not only how good an explanation it would be, but also what as yet unobserved contrasts it would explain, and this directs future observation and experiment. Semmelweis's research also shows that Inference to the Best Explanation is well suited to describe the role of overarching hypotheses in directing inquiry. Semmelweis's path to his cadaveric hypothesis is guided by his prior conjecture that the contrast in mortalities between the divisions is somehow due to the fact that deliveries are performed by medical students in the First Division, but by midwives in the Second Division. He then searches for ways of fleshing out this explanation and for the data that would test various proposals. Again, I have suggested that we can understand Semmelweis's rejection of the priest and the birth position hypotheses in terms of an inference to a negative explanation. The best explanation for the observed fact that eliminating these differences between the divisions did not affect mortality is that the mortality had a different cause. In both cases, the intermediate explanations focus research, either by marking the causal region within which the final explanation is likely to be found, or by showing that a certain region is unlikely to include the cause Semmelweis is looking for.

The hypothetico-deductive model emphasizes the priority of theory as a guide to observation and experiment, at the cost of neglecting the sources of theory. I want now to argue that the model also fails to give a good account of the way scientists decide which observations and experiments are worth making. According to the deductive model, scientists should check the observable consequences of their theoretical conjectures, or of their theoretical systems, consisting of the conjunction of theories and suitable auxiliary statements. As we will see below, this account is too restrictive, since there are relevant data not entailed by the theoretical system. It is also too permissive, since most consequences are not worth checking. Any hypothesis entails the disjunction of itself and any observational claim whatever, but establishing the truth of such a disjunction by checking the observable disjunct rarely bears on the truth of the hypothesis. The contrastive account of Inference to the Best Explanation is more informative, since it suggests Semmelweis's strategy of looking for observable contrasts that distinguish one causal hypothesis from competing explanations.

Even if we take both Semmelweis's hypotheses and his data as given, the hypothetico-deductive model gives a relatively poor account of their relevance to each other. This is particularly clear in the case of negative evidence. According to the deductive model, evidence disconfirms a hypothesis just in case the evidence either contradicts the hypothesis or contradicts the conjunction of the hypothesis and suitable auxiliary statements. None of the hypotheses Semmelweis rejects contradicts his data outright. For example, the epidemic hypothesis does not contradict the observed contrast in mortality between the divisions. Proponents of the epidemic hypothesis would have acknowledged that, like any other epidemic, not everyone who is exposed to the influence succumbs to the fever. They realized that not all mothers contract childbed fever, but rightly held that this did not refute their hypothesis, which was that the epidemic influence was a cause of the fever in those mothers that did contract it. So the hypothesis does not entail that the mortality in the two divisions is the same. Similarly, the delivery position hypothesis does not entail that the mortality in the two divisions is different when the birth positions are different; nor does it entail that the mortality will be the same when the positions are the same. Even if back delivery is a cause of childbed fever, the mortality in the Second Division could have been as high as in the First, because the fever might have had other causes there. Similarly, the possibility of additional causes shows that back delivery could be a cause of fever even though the mortality in the First Division is lower than in the Second Division when all the mothers deliver on their sides. The situation is the same for all the other hypotheses Semmelweis rejects.

What does Hempel say about this? He finds a logical conflict, but in the wrong place. According to him, the hypotheses that appealed to overcrowding, diet, or general care were rejected because the claims that the difference in mortality between the divisions was due to such differences 'conflict with readily observable facts' (1966, p. 6). The claim that, for example, the difference in mortality is due to a difference in diet is incompatible with the observation that there is no difference in diet. These are clearly cases of logical incompatibility, but they are not the ones Hempel needs: the claims that are incompatible with observation are not the general hypotheses Semmelweis rejects. Like the cadaveric hypothesis he eventually accepts, the hypotheses of overcrowding, diet, and care are surely general conjectures about causes of childbed fever, not specific claims about the differences between the divisions. But the hypotheses that overcrowding, diet, or general care is a cause of childbed fever are logically compatible with everything Semmelweis observes.

The hypothetico-deductivist must claim that hypotheses are rejected because, although they are compatible with the data, each of them, when conjoined with suitable auxiliary statements, is not. But what could such statements be? Each hypothesis must have a set of auxiliaries that allows the deduction that the mortality in the divisions is the same, and this contradicts the data. The auxiliaries need not be known to be true, but they need to be specified. This, however, cannot be done. The proponent of the epidemic hypothesis, for example, does not know what additional factors determine just who succumbs to the influence, so he cannot say how the divisions must be similar in order for it to follow that the mortality should be the same. Similarly, Semmelweis knew from the beginning that back delivery cannot be necessary for childbed fever, since there are cases of fever in the Second Division where all the women are delivered on their sides, but he cannot specify what all the other relevant factors ought to be. The best the hypothetico-deductivist can do, then, is to rely on *ceteris paribus* auxiliaries. If fever is caused by epidemic influence, or by back delivery, and everything else 'is equal', the mortality in the divisions ought to be the same. This, however, does not provide a useful analysis of the situation. Any proponent of the rejected hypotheses will reasonably claim that precisely what the contrast between the divisions shows is that not everything is equal. This shows that there is more to be said about the etiology of childbed fever, but it does not show why we should reject any of the hypotheses that Semmelweis does reject. Semmelweis's observations show that none of these hypotheses would explain the contrasts, but they do not show that the hypotheses are false, on hypothetico-deductive grounds.

There is another objection to the *ceteris paribus* approach, and indeed to any other scheme that would generate the auxiliaries the deductive model requires to account for negative evidence. It would disprove too much. Recall that the cadaveric hypothesis does not itself explain all the relevant contrasts, such as why some women in the Second Division contracted childbed fever while others in that division did not, or why some women who had 'street births' on their way to the hospital contracted the fever while others did not. If the other hypotheses were rejected because they, along with *ceteris paribus* clauses, entailed that there ought to be no difference in mortality between the divisions, then the model does not help us to understand why similar clauses did not lead Semmelweis to reject the cadaveric hypothesis as well.

From the point of view of Inference to the Best Explanation, we can see that there are several general and related reasons why the hypothetico-deductive model does not give a good description of the way causal hypotheses are disconfirmed by contrastive data. The most important is that the model does not account for the negative impact of explanatory failure. Semmelweis rejected hypotheses because they failed to explain contrasts, not because they were logically incompatible with them. Even on a deductive-nomological account of explanation, the failure to explain is not tantamount to a contradiction. In order to register the negative impact of these failures, the hypothetico-deductive model must place them on the Procrustean bed of logical incompatibility, which requires auxiliary statements that are not used by scientists and not usually available even if they were wanted.

Second, the hypothetico-deductive model misconstrues the nature of explanatory failure, in the case of contrastive explanations. As we saw in chapter three, to explain a contrast is not to deduce the conjunction of the fact and the negation of the foil, but to find some causal difference. The hypotheses Semmelweis rejects do not fail to explain because they do not entail the contrast between the divisions: the cadaveric hypothesis does not entail this either. They fail because they do not mark a difference between the divisions, either initially or after manipulation.

Third, the model does not reflect the need, in the case of explanatory failure, to judge whether this is due to incompleteness or error. In the model, this decision becomes one of whether we should reject the hypothesis or the auxiliaries in a case where their conjunction contradicts the evidence. This, however, is not the decision Semmelweis had to make. When he had all the mothers in both divisions deliver on their sides, and found that this did not affect the contrast in mortality, he did not have to choose between saying that the hypothesis that delivery position is a cause of fever is false and saying that the claim that everything else was equal is false. After his experiment, he knew that not everything else was equal, but this left him with the question of whether he ought to reject the delivery hypothesis or just judge it to be incomplete.

The failures of the hypothetico-deductive model to capture the force of disconfirmation through explanatory failure also clearly count against Karl Popper's account of theory testing through falsification (Popper, 1959). Although he is wrong to suppose that we can give an adequate account of science without relying on some notion of positive inductive support, Popper is right to suppose that much scientific research consists in the attempt to select from among competing conjectures by disconfirming all but one of them. Popper's mistake here is to hold that disconfirmation works exclusively through refutation. As the Semmelweis example shows, scientists also reject theories as false because, while they are not refuted by the evidence, they fail to explain the salient contrasts. Moreover, if my account of the way this sort of negative evidence operates is along the right lines, this is a form of disconfirmation that Popper's account cannot be modified to capture without abandoning his central proscription on positive support, since it requires that we make a positive judgment about whether the explanatory failure is more likely to be due to incompleteness or error, a judgment that depends on inductive considerations.

The hypothetico-deductive model appears to do a better job of accounting for Semmelweis's main positive argument for his cadaveric hypothesis, that disinfection eliminated the contrast in mortality between the divisions. Suppose we take it that the cadaveric hypothesis says that infection with cadaveric matter is a necessary cause of childbed fever, that everyone who contracts the fever was so infected. In this case, the hypothesis entails that where there is no infection, there is no fever. This, along with plausible auxiliaries about the influence of the disinfectant, entails that there should be no fever in the First Division after disinfection. But this analysis does not do justice to the experiment, for three reasons. First of all, the claim that cadaveric infection is strictly necessary for fever, which is needed for the deduction, is not strictly a tenable form of the cadaveric hypothesis, since Semmelweis knew of some cases of fever, such as those in the Second Division and those among street births, where there was no cadaveric infection. Similarly, given that disinfection is completely effective, this version of the hypothesis entails that there should be no cases of fever in the First Division after disinfection, which is not what Semmelweis observed. What he found was rather that the mortality in the First Division went down to the same low level (just over one per cent) as in the Second Division. As Hempel himself observes, Semmelweis eventually went on to 'broaden' his hypothesis, by allowing that childbed fever could also be caused by 'putrid matter derived from living organisms' (1966, p. 6). But if this is to count as broadening the hypothesis, rather than rejecting it, the original cadaveric hypothesis cannot have been that cadaveric infection is a necessary cause of the fever.

The second reason the deductive analysis of the disinfection experiment does not do justice to it is that the analysis does not bring out the special probative force of the contrastive experiment. Even if we suppose that cadaveric infection is necessary for fever, the hypothesis does not entail the *change* in mortality, but only that there should be no fever where there is disinfection, since it does not entail that there should be fever where there is no disinfection. But it is precisely this contrast that makes the experiment persuasive. What the hypothetico-deductivist could say here, I suppose, is that the change is entailed if we use the observed prior mortality as a premise in the argument. If cadaveric infection is necessary for fever, and if there was fever and infection but then the infection is removed, it follows that the fever will disappear as well. Even this, however, leaves out an essential feature of the experiment, which was the knowledge that, apart from disinfection, all the antecedents of the diachronic fact and foil were held constant. Finally, what makes the cadaveric experiment so telling is not only that it provides evidence that is well explained by the cadaveric hypothesis, but that the evidence simultaneously disconfirms the competitors. None of the other hypotheses can explain the temporal difference, since they all appeal to factors that were unchanged in this experiment. As we have seen in our discussion of negative evidence, however, the deductive model does not account for this process of disconfirmation through explanatory failure, and so it does not account for the way the evidence makes the cadaveric hypothesis the best explanation by simultaneously strengthening it and weakening its rivals.

I conclude that Inference to the Best Explanation, linked up to an account of contrastive explanation that provides an alternative to the deductive-nomological model, is an improvement over the hypothetico-deductive model in its account of the context of discovery, the determination of relevant evidence, the nature of disconfirmation, and the special positive support that certain contrastive experiments provide. In particular, Inference to the Best Explanation is an improvement because it allows for evidential relevance in the absence of plausible deductive connections, since contrastive explanations need not entail what they explain. If Inference to the Best Explanation is to come out as a suitable replacement for the hypothetico-deductive model, however, it is important to see that it does not conflict with the obvious fact that scientific research is shot through with deductive inferences. To deny that all scientific explanations can be cast in deductive form is not to deny that some of them can, or that deduction often plays an essential role in those that cannot be so cast. Semmelweis certainly relied on deductive inferences, many of them elementary. For example, he needed to use deductive calculations to determine the relative frequencies of fever mortalities for the two divisions and for street births.

Moreover, in many cases of causal scientific explanation, deduction is required to see whether a putative cause would explain a particular contrast. One reason for this is that an effect may be due to many causes, some of which are already known, and calculation is required to determine whether an additional putative cause would explain the residual effect. Consider, for example, the inference from the perturbation in the orbit of Uranus to the existence of Neptune. In order to determine whether Neptune would explain this perturbation, Adams and Leverrier had first to calculate the influence of the Sun and known planets on Uranus, in order to work out what the perturbation was, and then had to do further calculations to determine what sort of planet and orbit would account for it. As Mill points out, this 'Method of Residues' is an elaboration of the Method of Difference where the negative instance is 'not the direct result of observation and experiment, but has been arrived at by deduction' (1904, III.VIII.5). Through deduction, Adams and Leverrier determined that Neptune would explain why Uranus had a perturbed orbit rather than the one it would have had if only the Sun and known planets were influencing its motion. This example also illustrates other roles for deduction, since calculation was required to solve Newton's equations even for the sole influence of the Sun, and to go from the subsequent observations of Neptune to Neptune's mass and orbit. This particular inference to the best contrastive explanation would not have been possible without deduction.

Let us return now to the two challenges for Inference to the Best Explanation, that it mark an improvement over the hypothetico-deductive model, and that it tell us more than that inductive inference is often Inference to the Likeliest Cause. I have argued that the Semmelweis case shows that Inference to the Best Explanation passes the first test. It helps to show how the account passes the second test, by illustrating some of the ways explanatory considerations guide inference and judgments of likeliness. Although Semmelweis's overriding interest was in control rather than in understanding, he focused his inquiry by asking a contrastive explanatory question. Faced with the brute fact that many women were dying of childbed fever, and the many competing explanations for this, Semmelweis did not simply consider which explanation seemed the most plausible. Instead, he followed an organized research program based on evidential contrasts. By means of a combination of conjecture, observation, and manipulation, Semmelweis tried to show that the cadaveric hypothesis is the only available hypothesis that adequately explains his central contrast in mortality between the divisions. This entire process is governed by explanatory considerations that are not simply reducible to independent judgments of likeliness. By asking why the mortality in the two divisions was different, Semmelweis was able to generate a pool of candidate hypotheses, which he then evaluated by appeal to what they could and could not explain; his experimental procedure was governed by the need to find contrasts that would distinguish between them on explanatory grounds. When Semmelweis inferred the cadaveric hypothesis, it was not simply that what turned out to be the likeliest hypothesis also seemed the best explanation: he judged that the likeliest cause of most of the cases of childbed fever in his hospital was infection by cadaveric matter *because* this was the best explanation of his evidence.

The picture of Inference to the Best Explanation that has emerged from the example of Semmelweis's research is, I think, somewhat different from the one that the slogan initially suggests in two important respects. The slogan calls to mind a fairly passive process, where we take whatever data happen to be to hand and infer an explanation, and where the central judgment we must make in this process is which of a battery of explanations of the same data would, if true, provide the loveliest explanation. But as the example shows, Inference to the Best Explanation supports a picture of research that is at once more active and realistic, where explanatory considerations guide the program of observation and experiment, as well as of conjecture. The upshot of this program is an inference to the loveliest explanation but the technique is eliminative. Through the use of judiciously chosen experiments, Semmelweis determined the loveliest explanation by a process of manipulation and elimination that left only a single explanation of the salient contrasts. In effect, he converted the question of the loveliest explanation of non-contrastive facts into the question of the only explanation of various contrasts. Research programs that make this conversion are common in science, and it is one of the merits of Inference to the Best Explanation that it elucidates this strategy. And it is because Semmelweis successfully pursued it that we have been able to say something substantial about how explanatory considerations can be a guide to inference without getting bogged down in the daunting question of comparative loveliness where two hypotheses do both explain the same data. At the same time, this question cannot be avoided in a full assessment of Inference to the Best Explanation, since scientists are not always as fortunate as Semmelweis was in finding contrasts that discriminate between all the competitors. Accordingly, I will attempt a very partial answer to it in chapter seven. First, however, I will consider the resources of Inference to the Best Explanation to avoid some of the over-permissiveness of the hypothetico-deductive model.

6

THE RAVEN PARADOX

UNSUITABLE CONTRASTS

The hypothetico-deductive model has three weaknesses. It neglects the context of discovery; it is too strict, discounting relevant evidence that is either not entailed by the hypothesis under investigation or not incompatible with it; and it is over-permissive, counting some irrelevant data as relevant. In the last chapter, I argued that Inference to the Best Explanation does better in the first two respects. In this chapter, I argue that it also does better in the third. The problem of over-permissiveness arises because there are consequences of a hypothetical system that do not support the hypothesis, and there are several related ways to generate such consequences. One is by strengthening the premise set. An observed consequence of a hypothesis about childbed fever may support that hypothesis, but it will not support the conjunctive hypothesis consisting of the fever hypothesis and some unrelated hypothesis, even though the conjunction will of course also entail the observation (cf. Goodman, 1983, pp. 67–8). Similar problems arise with the indiscriminate use of auxiliary statements. At the limit, any hypothesis is supported by any datum consistent with it, if we use a conditional auxiliary whose antecedent is the hypothesis and whose consequent is the irrelevant datum. This places the hypothetico-deductivist in a bind: to meet the objections in the last chapter that some supporting evidence is not entailed by the hypothesis, he will have to be extremely permissive about the sort of auxiliaries he allows, but what he then gains in coverage he pays for by also admitting irrelevant evidence.

A second way to generate irrelevant consequences is to

weaken the conclusion. For any consequence of a hypothesis that supports the hypothesis, there are also innumerable disjunctions of that consequence and the description of an irrelevant observation, and the truth of the disjunction can then be established by the irrelevant observation. 'All sodium burns yellow' entails that either this piece of sodium will burn yellow or there is a pen on my table, but the observation of the pen is irrelevant to the hypothesis. Third, as Goodman has shown, by using factitious predicates we can construct hypotheses that are not lawlike, which is to say that they are not supported by the instances they entail (Goodman, 1983, pp. 72–5). A grue emerald observed today does not support the hypothesis that all emeralds are grue, where something is grue just in case it is either observed before midnight tonight and green or not so observed and blue. Finally, we may generate apparently irrelevant consequences through contraposition: this is Hempel's raven paradox. The hypothesis that all F's are G entails that this F is G, a consequence that seems to support the hypothesis. Since, however, this hypothesis is logically equivalent to 'All non-G's are non-F', it also entails that this non-G is non-F, a consequence that does not seem to support the original hypothesis. 'All ravens are black' is logically equivalent to 'All non-black things are non-ravens', so the raven hypothesis entails that this white object (my shoe) is not a raven, but observations of my shoe seem irrelevant to the hypothesis (Hempel, 1965, pp. 14–15).

Inference to the Best Explanation has the resources to avoid some of these irrelevant consequences. One way of avoiding the problem of arbitrary conjuncts is to observe that a causal account of explanation, unlike the deductive-nomological model, does not allow such arbitrary strengthening. Adding irrelevant information to a good explanation can ruin it. On a causal model of explanation, when we say that something happened *because* of X, we are claiming that X is a cause, and this claim will be false if X includes causally irrelevant material. Alternatively, one might argue that logically strengthening the explanation by adding a gratuitous conjunct worsens the explanation, even if it does not spoil it altogether, so we ought only to infer the weaker premise. Similarly, Inference to the Best Explanation can rule out some weakened consequences, since the sodium hypothesis does not explain the disjunction of itself and an arbitrary statement. It is less clear what to say in the case of the disjunction of a relevant instance and an irrelevant one. Perhaps arbitrary disjunctions are not suitable objects of causal explanation, since it is unclear whether it makes sense to ask for the cause of such disjunctions. As for Goodman's new riddle of induction, it may be argued that while 'All emeralds are green' enables us to give an obliquely causal explanation of why this object is green, 'All emeralds are grue' does not explain why an object is grue, because 'grue' does not pick out the sort of feature that can be an effect. Finally, explanatory considerations seem to help with the raven paradox since, while the raven hypothesis may provide some sort of explanation for the blackness of a particular raven, neither the original hypothesis nor its contrapositive explain why this shoe is white. Inductive irrelevance and explanatory irrelevance appear largely to coincide.

All of these suggestions would require development and defense, but in what follows I will concentrate only on the case of the raven paradox, since it is here that my account of contrastive inference gets the best purchase. My main conclusion will be that the raven paradox does not arise for contrastive inferences. It will be useful to begin by considering briefly three well-known solutions to the raven paradox, which will provide foils to my approach. The first is Hempel's own strategy, which is to bite the bullet. He frames the question of whether observing a non-black non-raven supports the raven hypothesis with the 'methodo-logical fiction' that this observation is the only available information that might be relevant to the hypothesis; that is, that there is no relevant background information or other relevant evidence. He then claims that, under this idealizing assumption, a non-black non-raven does indeed provide some support for the hypothesis that all ravens are black, and that any intuitions to the contrary are due to the failure to respect the idealization (Hempel, 1965, pp. 18–20). This solution is consistent with the hypothetico-deductive model, but it is also unsatisfying, for three reasons. First of all, even if white shoes do support the raven hypothesis under the idealization, this leaves the interesting question unanswered, which is why in methodological fact we do not look to non-black non-ravens for support of the raven hypothesis. Second, the methodological fiction takes us so far from our actual inductive practice that there may be no way to determine what the right answer in the fictional case ought to be. Perhaps it is instead that even black ravens would not support the hypothesis under such extreme ignorance since, if we really knew nothing else relevant to the hypothesis, we would have no reason to believe that it is lawlike, that is, confirmable even by its direct instances. Not all generalizations are instance confirmable so, if we do not know anything relevant to the raven hypothesis, except that there exists one black raven, or one non-black non-raven, we do not know whether the hypothesis is confirmable or not. Finally, it is not clear that the idealization is sensible, because it is not clear that the question of support is coherent under those conditions. It makes sense to ask what the period of a pendulum is under the idealization of no friction, but not under the idealization of no length since, without length, you have no pendulum (cf. Cummins, 1989, p. 78). Similarly, there may be no such thing as inductive support without background information, just as there is no such thing as support without a hypothesis to be supported.

Another well-known attempt to solve the problem is due to Quine who, drawing on Goodman's discussion of the new riddle of induction, suggests that only instances of 'projectible' predicates provide inductive support and that the complement of a projectible predicate is not projectible. On this view, black ravens support both the raven hypothesis and its contrapositive, but non-black non-ravens support neither (Quine, 1969, pp. 114–16). This solution is not compatible with the hypothetico-deductive model, since it makes many consequences of a hypothesis inductively irrelevant to it. The main objection to this proposal is that some complements of projectible predicates are projectible. For example, some things that are neither rubbed nor heated do support the hypothesis that friction causes heat. As we will shortly see, this is far from an isolated case.

The last solution I will mention is Goodman's own. He claims that inductive support can be analyzed in terms of 'selective confirmation', which requires that confirming evidence both provide an instance of the hypothesis and a counterexample to its contrary. The contrary of 'All ravens are black' is 'No Ravens are black', which is incompatible with 'This raven is black', but not with 'This shoe is white' (Goodman, 1983, pp. 70–1). Like Quine's proposal, this is incompatible with simple hypothetico-deductivism. The most serious objection to Goodman's solution is that, unlike Quine's proposal, it violates the very plausible equivalence condition on inductive support, that whatever supports a hypothesis also supports any logically equivalent hypothesis. A white shoe does not selectively confirm the raven hypothesis, but it does selectively confirm the logically equivalent hypothesis that all non-black things are non-ravens, since it is incompatible with the contrary hypothesis that no non-black things are non-ravens. It hardly removes the sense of paradox to be told that, while white shoes do not support the raven hypothesis, they do support a logical equivalent.

Bearing these proposals and their drawbacks in mind, I want now to argue that contrastive inference avoids the paradox. To do so, it is best to leave the particular example of the ravens to one side for the moment and to consider instead a more straightforwardly causal example. We will then be in a better position to take care of those puzzling birds. In spite of its apparent simplicity, the raven hypothesis raises special problems, and an excessive focus on this example has made the general problem seem harder than it is. Semmelweis's cadaveric hypothesis, which so occupied our attention in the last chapter, is a more suitable case. The hypothesis is that childbed fever is caused by cadaveric infection. Simplifying somewhat, let us suppose that Semmelweis's hypothesis has the same form as the raven hypothesis, that all cases of mothers with cadaveric infection are cases of mothers with childbed fever.

Semmelweis's evidence for the cadaveric hypothesis consisted primarily in two contrasts, one synchronic and one diachronic. He noticed that the hypothesis would explain why the incidence of the fever was much higher in one maternity division of the hospital than in the other at the same time, since it was only in the First Division that mothers were examined by people who had handled cadaveric matter. It would also explain why the incidence of fever in the First Division went down to the same low level as the Second Division after he had the medics who performed the examinations in the First Division disinfect their hands. In each case, Semmelweis relied *both* on instances of infection and fever and on instances of non-fever and non-infection, that is, both on instances of the direct hypothesis that all infections are fevers and on instances of the equivalent contrapositive hypothesis that all non-fevers are non-infections.

Contrastive inferences place the raven paradox in a new light. The problem is no longer to show why contrapositive instances do not provide support, or to explain away the intuition that they

103

do not. In contrastive inferences, they provide strong support and this is consonant with our intuitions about these cases. Contrapositive instances are essential to contrastive inference (or to the Method of Difference, upon which much of my account of contrastive inference has so far been based), itself an indispensable vehicle of inductive support. The real problem is rather to show why some contrapositive instances support while others do not. The mothers who were not infected and did not contract the fever were contrapositive instances that provided an essential part of Semmelweis's evidence for his hypothesis, but Semmelweis's observation that his shoe was neither infected nor feverish would not have supported his hypothesis.

The structure of contrastive inference suggests a natural solution to this problem. The uninfected mothers supported Semmelweis's hypothesis because they provided a suitable foil to the direct instances of the infected mothers with fever. As we have seen, the main condition for the suitability of a foil is shared history with its fact. The ideal case is one where the only difference in the history of fact and foil is the presence of the putative cause. This ideal can never be strictly achieved, but the investigator will look for contrasts that appear to differ in no respects that are likely to be causally relevant, or at least for contrasts that share other suspect causes. One reason Semmelweis's shoe did not support his hypothesis is that it provided a contrast that was too different from infected mothers, so there was no reason to say that those mothers contracted the fever while the shoe did not because only the mothers were infected by cadaveric matter. Infection cannot be inferred as the best explanation of this contrast, because there are too many other differences.

The requirement that a supporting contrapositive instance be known to share most of its history with an observed direct instance shows how background knowledge is essential to contrastive support. Unless we have this knowledge, we cannot make the inference. Nobody would suggest that we consider applications of the Method of Difference under the methodological fiction that we know nothing about the antecedents of fact and foil, since the method simply would not apply to such a case. Background knowledge may rule out a contrapositive instance either because we do not know enough about the histories of fact and foil or because we do know that they differ in many respects.

Another way background knowledge may rule out contrastive inference is if we already know that the absence of the effect in the case of the foil is not due to the absence of the putative cause. In the case of the shoe, Semmelweis already knew that, even if cadaveric infection was indeed a cause of childbed fever, the reason his shoe did not have fever was not because it was uninfected. He knew that only living organisms can contract fever, and this pre-empted the explanation by appeal to lack of infection. A foil is unsuitable if it is already known that it would not have manifested the effect, even if it had the putative cause. In sum, then, a contrapositive instance only provides support if it is known to have a similar history to a direct instance and if it is not already known that the absence of the effect is due to some pre-empting cause. It is only if both of these conditions are satisfied that the hypothesis is a candidate for the best explanation of the contrast.

The method of contrastive inference avoids the raven paradox, because the restrictions on suitable contrasts rule out irrelevant instances. This is a solution to the paradox that differs from those of Hempel, Quine, and Goodman. Unlike Hempel's solution, it allows for a distinction between relevant and irrelevant instances, though this does not line up with the distinction between direct and contrapositive instances, and the contrastive solution does not require that we fly in the face of our intuitions about inductive relevance. It also rejects the terms of methodological fiction Hempel uses to frame his position, since the question of whether a contrapositive instance is a suitable foil can only be answered by appeal to background knowledge. Finally, since the contrastive inference does not count all contrapositive instances as providing support it is, unlike Hempel's solution, incompatible with the hypothetico-deductive model. (As we saw in the last chapter, this incompatibility holds for other reasons as well.) The contrastive solution is also different from Quine's appeal to projectibility, since it shows that the complement of a projectible predicate may itself be projectible, and indeed must be in any situation where contrastive inferences can be made. Lastly, the contrastive solution differs from Goodman's use of selective confirmation, since it does not violate the equivalence condition. A hypothesis and its contrapositive are supported by just the same contrastive data. Whichever form of hypothesis we use, we need the same pairs of instances subject to the same constraints of similar histories and the absence of known pre-empting explanations.

This completes my central argument for the claim that Inference to the Best Explanation avoids some of the over-permissiveness of the deductive-nomological model. Not all contrapositive instances can be used to infer a contrastive explanation, and the model helps to say which can and which cannot. But I suppose that no discussion of the raven paradox can ignore the ravens themselves, so we will now return to them, though what I will have to say about them is not essential to my general claim. Contrastive inference does not suffer from the raven paradox, but neither does it provide much help with the particular example of the raven hypothesis, that all ravens are black. The trouble is that this hypothesis does not find support in contrastive evidence at all. To find such evidence, we might look for birds similar to ravens that are not black that might be paired up with observed black ravens, but in fact this is not something worth doing. The problem is not that there are no such birds, but that finding them would not support the raven hypothesis. If we found those birds, they might even tend to disconfirm the raven hypothesis: if there are non-black birds very much like ravens in other respects, this may increase the likelihood that there are some non-black ravens we have yet to observe.

Why can the raven hypothesis not find contrastive support? One natural suspicion is that this is because it is not a causal hypothesis. This, however, is not the problem. First, the hypothesis is broadly causal, though the cause is only obliquely described. Here is one way of analyzing the situation. When we consider the hypothesis that all ravens are black, what we are considering is that there is something in ravens, a gene perhaps, that makes them black. Moreover, the hypothesis implies that this cause, whatever it is, is part of the essence of ravens (or at least nomologically linked to their essence). Roughly speaking, the hypothesis implies that birds similar to ravens except with respect to this cause would not interbreed with ravens. The hypothesis thus implicitly makes a claim that could be false even as it applies to ravens that are in fact black, since the cause of blackness could be merely accidental to those birds, a feature they could lack without forfeiting their species membership. The raven hypothesis is causal, but the cause is only characterized as something in ravens that is essential to them. This characterization is very broad, but it is necessary if we are to make sense of the way we would support the hypothesis. For if we supposed instead that the hypothesis is entirely neutral with respect to the essentiality of the cause, it would not be lawlike; that is, it would not be supported even by black ravens. If we thought it was merely some contingent feature of the black ravens we observed that made them black, then even observing many black ravens should give us no confidence that all ravens, past, present, and future, are black. This raises the question of how we could experimentally distinguish between an essential and accidental cause, a question we will address in the next section.

The reason the raven hypothesis is not susceptible to contrastive support cannot be that it is non-causal, if in fact it is causal. But even if its causal status is questionable, there is reason to believe that this is not the source of the problem, since contrastive inference is applicable to other hypotheses that have no stronger claim to causal status. Consider the hypothesis that all sodium burns yellow. Being sodium does not cause something to have the dispositional property of burning yellow any more or less than being a raven causes it to be black. As in the case of the raven hypothesis, I would say that the sodium hypothesis claims that there is something in sodium that causes it to burn yellow, and that this feature is essential to sodium. So the sodium hypothesis does not have a stronger claim to causal status than the raven hypothesis, yet it is contrastively supported. One way we convince ourselves of its truth is by producing a flame that has no sodium in it and is not burning yellow, and then adding sodium and noticing the immediate change to a yellow flame. This diachronic contrast clearly supports the hypothesis.

What then is the difference between the sodium hypothesis and the raven hypothesis that explains why one enjoys contrastive support and the other does not? It is easy to see one reason why the raven hypothesis does not lend itself to diachronic contrast. We cannot transform a non-black non-raven into a raven to see whether we get a simultaneous transformation from non-black to black, in the way that we can transform a flame without sodium into a flame with sodium. Perhaps if such a transformation were possible, we could usefully apply the Method of Difference to the raven hypothesis. There is probably another reason why diachronic contrast is useful for sodium, but not for ravens. In the case of yellow flames that contain sodium, we may wonder whether it is the sodium or some other factor that is responsible for the color. The diachronic contrast eliminates other suspects. In the case of the ravens, on the other hand, we already presume that the color is caused by something in the ravens rather than, say, the lighting conditions under which ravens are generally observed, so we do not look to experiments that would support this. For the raven hypothesis, only the essentiality claim and not the causal claim are in question. For both these reasons, then, it is no mystery that the raven hypothesis is not susceptible to diachronic contrastive support. What about the synchronic case? Here the sodium and the ravens are similar: neither finds support. We have already seen this for the ravens. Similarly, even if there are elements like sodium that do not burn yellow, observing this contrast would not increase our confidence that all sodium burns yellow.

Neither hypothesis lends itself to synchronic contrastive support, as a matter of principle, and the raven hypothesis does not lend itself to a diachronic application, as a matter of fact, since we do not know how to transform non-ravens into ravens and we do already know that the ravens themselves are responsible for their color. The remaining question, then, is why there is not synchronic application in either case. The answer, I think, lies in the extreme obliqueness of the causal descriptions implicit in the two hypotheses. This makes it impossible for us to know, in the synchronic case, whether we have satisfied the shared background condition. Faced with a raven and a generally similar but non-black bird, we know that the cause of the blackness of the raven must lie among the differences, but we are none the wiser about what the relevant difference is, and whether it is essential or accidental to ravens, the crucial question. Similar remarks apply to a contrast between a sample of sodium and an apparently similar substance that does not burn yellow.

My view that it is the obliqueness of the causal description that disqualifies the raven hypothesis from synchronic contrastive support is corroborated by comparing this case to a more directly causal hypothesis about the coloration of birds. Suppose we wanted to test the hypothesis that gene B causes blackness, where we have some independent way to identify the presence of that gene. In this case, finding that all the black birds we have observed have gene B, and that otherwise genetically similar birds without gene B are uniformly non-black, would support the genetic hypothesis. Without something like the genetic hypothesis, however, we simply do not know what should count as a relevantly similar bird. Similar considerations apply to the case of sodium. If we had a hypothesis about just what feature of sodium causes it to burn yellow, and we found an element that was just like sodium except with respect to this feature and burning color, we would have a useful synchronic contrast. Without this, we do not know what a relevantly similar element would be. In the case of sodium, we can avoid this problem by using a diachronic contrast but, as we have seen, this is not an option for ravens.

THE METHOD OF AGREEMENT

Contrastive inference avoids the raven paradox, but it does not account for the way we support the raven hypothesis itself. Since contrastive inference is modeled on Mill's Method of Difference, the natural place to look for a kind of causal inference that would cover the ravens is in Mill's other main method, the Method of Agreement. In the Method of Agreement, we look for two instances of the effect whose histories ideally have nothing in common but the putative cause (Mill, 1904, III.VIII.1). Here we hold the effect constant and vary the history to see what stays the same, while in contrastive inference or the Method of Difference, we hold the history constant and vary the effect. The inference to the cadaveric hypothesis from the evidence of infected mothers who contracted fever, even though they differed in birth position, exposure to the priest, diet, and so on, would be an application of the Method of Agreement.

Like the Method of Difference, applications of the Method of Agreement are naturally construed as inferences to the best explanation, where what is explained is a similarity rather than a contrast. To explain a similarity, we must cite a cause of the effect shared by the instances. For example, to explain why all the members of the club have red hair, we might point out that red hair is a requirement for admission. This explains why all of them have red hair, even though it does not explain why each of them does (cf. Nozick, 1974, p. 22), so explanations of similarity exhibit a divergence between explaining P and explaining 'P and Q', like the divergence we found in chapter three between explaining P and explaining 'P rather than Q'. Similarly, the cause that we cite to explain P in one milieu of similar effects may differ as we change the milieu, since a common cause in one case may not be in another. In the context of inference, the evidence of agreement focuses our inference on these shared elements that would explain the agreement, much as the contrastive evidence focuses our inference on explanatory differences.

The primary restriction on supporting instances of agreement is varying history. Ideally the pair of instances share only effect and putative cause; in practice, what is required is that the instances must at least differ in the presence of something that might be a cause of the effect. The Method of Agreement requires a further restriction, because of the risk of what Mill calls 'the plurality of causes'. As he observes, the method may give the wrong result in cases where the effect in question can be caused in more than one way (III.X.2). Suppose that two people die, one from poisoning, the other from a knife wound. The Method of Agreement will discount both causes, since they are part of the variation between the instances. Worse, the method may count as cause an incidental similarity between the two, say that they both smoked. Mill suggests two ways of handling this problem. The first is also to apply the Method of Difference, where this is possible. The second is to gather additional and varied instances for the Method of Agreement: if many instances share only one suspect, either this is a cause or there are as many causes as there are instances, an unlikely result. There is, however, a third obvious way to handle the problem of the plurality of causes, and this imposes a further restriction on the instances suitable to the method. We will only use pairs of instances that our background knowledge leads us to believe have the same kind of cause, even if we do not yet know what it is.

These two restrictions of varied history and common etiology seem to protect the Method of Agreement from the raven paradox. Unlike contrastive inference, which always joins a direct instance and a contrapositive instance, the Method of Agreement requires different pairs for a hypothesis and its contrapositive. For the hypothesis that all infected mothers contract fever, we would look at a pair of infected mothers with fever; for the contrapositive that everything that does not contract childbed fever is not an infected mother, we might look instead at a pair of uninfected and healthy mothers. As in the case of contrastive inference, the Method of Agreement correctly allows that some contrapositive instances provide support, such as a pair of uninfected mothers without fever, who had different delivery positions. Similarly, if we wanted to know whether a vitamin C deficiency causes scurvy, we would find support among healthy people without the deficiency, as well as among sufferers of scurvy with it. The method also correctly excludes a pair of shoes in the case of the cadaveric hypothesis, since these shoes do not differ in any respect that we think might be causally relevant to contracting childbed fever. But what about a pair consisting of a healthy mother without fever and an uninfected shoe? The histories of the mother and the shoe vary a great deal, yet we do not want to say that they support the cadaveric hypothesis. I think we can exclude cases of this sort by appeal to the common etiology condition. We do not believe the shoe is free from childbed fever because it has not been infected, even if infection is the cause of fever; the reason shoes do not have fever is that they are not living organisms.

Now we can return to the ravens themselves. To determine whether a pair of agreeing instances would support the raven hypothesis, we must consider what the alternative causes might be. This is not obvious, because the raven hypothesis is only obliquely causal. Following my suggestion in the last section, let us take the raven hypothesis to imply that the cause of blackness in ravens is some feature of ravens that is essential to them, and the other suspects to be features of ravens that are non-essential. How can we discriminate empirically between these possibilities? The best we can do at the level of the raven hypothesis is to employ a direct application of the Method of Agreement. By finding only black ravens in varied locations and environments, we eliminate various accidental suspects, namely those that vary, and this provides some support for the hypothesis.

Why will we not bother to look for two very different non-black non-ravens? Some contrapositive instances are clearly irrelevant to the raven hypothesis, because they violate the common etiology restriction: the reason my shoe is white is not that it lacks some feature essential to ravens that makes them black. But why will we not bother to look at non-black birds? The reason is again the obliqueness of our description of the cause of blackness in ravens. If we observe one non-black bird, we know that it lacks whatever it is that makes black ravens black. But we are none the wiser about whether this missing factor is essential or accidental

111

to ravens, which is the point at issue. If we have a pair of very different non-black birds, we may infer that the missing factor is among those that they share. This narrows down the suspects, but we still have no way of saying whether the remaining candidates are essential or accidental to ravens. Indeed, even if we discovered that these birds are not black because they do not have gene B, the 'blackness gene', we are in no better shape, since this information does not help us to determine whether the presence of this gene is essential to ravens. This, I think, is the solution to the raven paradox as it applies to the peculiar example that gives it its name. When we look at various black ravens, at least we are ruling out some alternative causes, since we know that every respect in which these ravens differ is non-essential to them, but when we look at non-black birds, the information that we may gain about the etiology of coloration does us no good, since it does not discriminate between the raven hypothesis and its competitors.

In the last chapter, I argued that contrastive inference is a common evidential procedure that brings out many of the virtues of Inference to the Best Explanation and shows some of the ways it differs from the hypothetico-deductive model. In this chapter, I have continued that argument by suggesting why contrastive inference does not have to swallow the raven paradox. Since the hypothetico-deductive model does, this marks another difference that is to the credit of Inference to the Best Explanation. At the same time, the brief discussion of the Method of Agreement in this section underlines the fact that contrastive inference cannot be the whole of induction though, as I suggested, the Method of Agreement also seems amenable to explanatory analysis. I have chosen to focus in this book on the contrastive case, not because I think that it exhausts Inference to the Best Explanation, but because it is a particularly important form of inference and one that links up with a form of explanation that I have been able to articulate in some detail. What made this articulation possible is the close analogy between contrastive explanation and the Method of Difference. The analogy is so close that some readers, convinced that Inference to the Best Explanation really is different from hypothetico-deductivism, may now wonder whether it is not just a catchy name for Mill's method itself, rather than a distinctive account. This raises a third challenge to Inference to the Best Explanation. Even if the account is more than Inference to the Likeliest Cause, and improves on the hypothetico-deductive model, is it really any better than Mill's methods? In the next chapter, I consider this new challenge.

7

LOVELINESS AND INFERENCE

MULTIPLE DIFFERENCES

The Method of Difference has two serious limitations not shared by Inference to the Best Explanation. The first is that it does not account for inferred differences. The Method of Difference sanctions the inference that the only difference between the antecedents of a case where the effect occurs and one where it does not marks a cause of the effect. Here the contrastive evidence is not evidence for the *existence* of the prior difference, but only for its causal role. The method says nothing about the discovery of differences, only about the inference from sole difference to cause. So it does not describe the workings of the many contrastive inferences where the existence of the difference must be inferred, either because it is unobservable or because it is observable but not observed. Many of these cases are naturally described by Inference to the Best Explanation, since the difference is inferred precisely because it would explain the contrast.

Why does Inference to the Best Explanation account for inferred differences while the Method of Difference does not? The crucial point is that only Inference to the Best Explanation has us consider the connection between the putative cause and the evidence as a part of the process of inference. In an application of the Method of Difference, having found some contrastive evidence, we look at the antecedents or histories of the two cases. What governs our inference is not any relation between the histories and the evidence, but only a question about the histories themselves – whether they differ in only one respect. Inference to the Best Explanation works differently. Here we consider the potential explanatory connection between a difference and the contrastive evidence in order to determine what we should infer. We are to infer that a difference marks a cause just in case the difference would provide the best explanation of the contrast. Because of this subjunctive process, absent from the Method of Difference, we may judge that the difference that would best explain the evidence is one we do not observe, in which case Inference to the Best Explanation sanctions an inference to the existence of the difference, as well as to its causal role. Semmelweis did not observe that only women in the First Division were infected by cadaveric matter, so the Method of Difference does not show why he inferred the existence of this difference. Inference to the Best Explanation does show this, since the difference would explain the contrast in mortality between the divisions.

So the first advantage of Inference to the Best Explanation is that it extends the Method of Difference to inferred differences. Given how common such inferences are, this is a substantial advance; but it may still understate this case for the superiority of Inference to the Best Explanation. If the Method of Difference does not account for inferred differences, it is unclear how it can account for any contrastive inferences at all. The trouble lies in the requirement that we know that there is only one difference. Even if only one difference is observed, the method also requires that we judge that there are no unobserved differences, but it gives no account of the way this judgment is made.

The second weakness of the Method of Difference is that it does not account for the way we select from among multiple differences. Although Mill's strict statement of the Method of Difference sanctions an inference only when we know that there is a sole difference in the histories of fact and foil, Mill recognizes that this is an idealization. However similar the fact and foil, there will always be more than one difference between their antecedents. Some of these will be causally relevant, but others not. The problem of multiple differences is the problem of making this discrimination. Mill proposes two solutions. First, we may ignore differences 'such as are already known to be immaterial to the result' (III.VIII.3). Second, while passive observations will seldom satisfy the requirements of the method, Mill claims that carefully controlled experiments where a precise change is introduced into a system will often leave that change as the only possibly material difference between the situation before the change and the situation afterwards. Neither of these solutions is adequate. The first allows us to discount some differences, if we know they are irrelevant, but does not tell us how we determine irrelevance. The second shows that Mill is too sanguine about the powers of experimental technique to eliminate all but one possibly relevant difference. An account of contrastive inference should show how we select from among possibly relevant differences, something Mill's method does not do.

Mill does not seem to have realized the magnitude of the problem of multiple differences. One reason for this may be that he did not fully appreciate the need to take account of unobservable differences. Another is that he seems not to have seen that his own deterministic assumptions entail that the strict requirement of a single difference is one that can never be met, as a matter of principle. (I owe this point to Trevor Hussey.) Given any causal difference between the antecedents of fact and foil, there must, on Mill's assumptions, be a prior difference to account for that one, and for that one, and so on. There must always be multiple differences, so Mill can never say that a particular difference must be causally relevant on the grounds that it is the only difference.

The second advantage of Inference to the Best Explanation, I suggest, is that it elucidates the way we solve the problem of multiple differences, the way we judge whether a difference is a causal difference. Inference to the Best Explanation does not sanction a causal inference for any known or possible difference, but only for those that would provide the best (loveliest) explanation of the evidence. This answer, however, brings us back to the general and difficult problem of giving an account of what makes for a good explanation. The account of contrastive explanation in chapter three was a partial solution to this problem, and one that enabled us to see how explanatory considerations could be a guide to inference. Where we have contrastive data, Inference to the Best Explanation focuses our inferences on the relatively narrow class of possible causes that would also mark differences between the histories of a case where the effect occurs and a case where it does not, since only these differences would provide good explanations of the contrast. This explanatory consideration rules out the many possible causes that would not mark a difference between fact and foil. It does not, however, help much with the problem of multiple differences, which is just the problem of selecting from among the remaining candidates which do cite a difference. Inference to the Best Explanation, unlike the Method of Difference, can account for some of this process of selection, but this must depend on aspects of loveliness that go beyond the basic Difference Condition on contrastive explanation.

A detailed account of these further aspects of loveliness is beyond my present competence, but I hope in brief compass to make plausible the claim that explanatory considerations help to select those differences that are likely to be causes. The mechanism by which we settle on which of the many possible differences to infer has two stages. The first is the process of generation, the result of which is that we only consider a small portion of the possible differences; the second is the process of selection from among those live candidates. Explanatory considerations play a role in both stages.

Let us focus first on selection. The obvious way to select causes from differences left by a particular evidential contrast is to perform more experiments. This is what Semmelweis did. The initial contrast in mortality between the two maternity divisions might have been due to the difference in exposure to the priest, in birth position, or in infection by cadaveric matter. But Semmelweis found that only eliminating the third difference made a difference to the mortality rates, so he inferred that this difference marked a cause. Even if the three differences were equally good explanations of the initial contrast, the third was the best explanation of the total evidence. Semmelweis did not make his inference until he was able to make this discrimination. It is also easy to find examples where a hasty inference could have been avoided with additional data that would have ruled out an irrelevant difference, on explanatory grounds. In the seventeenth century, Sir Kenelm Digby, a founding member of the Royal Society, enthusiastically endorsed the idea, attributed to Paracelsus, that there was a 'sympathetic powder' that could cure wounds at a distance by, for example, rubbing it on the sword that caused the wound (Gjertsen, 1989, pp. 108–9). This hypothesis found contrastive support, since patients treated with sympathetic powder recovered more quickly than patients whose wounds were dressed by doctors and nurses in the normal way. The real reason for this contrast (we now suppose) was that the doctors and nurses inadvertently infected the wounds they were trying to heal, while the patients sympathetically treated were in effect left alone (only

117

the swords were treated), so were less likely to be infected. Additional data, comparing sympathetic treatment and no treatment at all, would have prevented the mistaken inference.

Among the differences that all the evidence leaves in the running, we prefer those that would mark causes we can link to the effect by some articulated causal mechanism and whose description allows us to deduce the precise details of the effect. Since we can specify a mechanism and make precise deductions from some differences, but not for others, these are criteria that sometimes help us to select causes from differences. And since mechanism and precision are explanatory virtues, Inference to the Best Explanation can account for these principles of selection. We understand a phenomenon better when we know, not just what caused it, but how the cause operated. And we understand more when we can explain the quantitative features of a phenomenon, and not just its qualitative ones. The importance of these explanatory virtues helps to bring out what is right about the deductive-nomological model, since giving an argument linking cause and effect can be a good way of spelling out a mechanism, and quantitative explanation is virtually impossible without deduction. This does not vindicate the deductive model, for reasons I have given in the last two chapters, but it does suggest that Inference to the Best Explanation has the resources to capture some of the most useful parts of both Mill's methods and a deductive model of theory evaluation and so is an improvement on both.

A preference for explanations that specify causal mechanisms and for explanations that give precise accounts of quantitative phenomena helps to reduce the problem of multiple differences, but this cannot be the whole story. We need to make a further appeal to the way our background beliefs help us to select differences. In mundane cases, we often have beliefs about the sort of causal mechanisms that are relevant to the case at hand, and this helps us to select from among differences. When my computer did not come on this morning, I at first inferred that there was a blown fuse, since this would explain the contrast between its behavior on other days and its lack of behavior today. Then I noticed another difference: my computer was unplugged. Given my background knowledge about electricity, I had little difficulty selecting which difference was the more likely cause. Even if the fuse had blown, this would not explain why my computer was not running.

Our background beliefs also influence our judgments of plausibility in more abstract ways. Take the case of the sympathetic powder. Even if we had no good data about the contrast between patients treated with sympathetic powder and patients left alone, we would not now infer that the powder explains the contrast between patients treated by powder and those treated by seventeenth-century doctors and nurses. The idea that the powder cures at a distance is just too much at variance with our other beliefs about the etiology of healing. On the other hand, the idea that powder at a distance is inert, but that somehow the doctors and nurses were inadvertently exacerbating the injury, fits well with other things we believe, in particular with the other explanations we already accept. This preference for familiar causes and mechanisms can be seen as a preference based in part on explanatory considerations. It is, in part, a preference for a unified explanatory scheme, and unification makes for lovely explanations (cf. Friedman, 1974).

These, then, are some of the ways Inference to the Best Explanation can account for the way we select causes from differences. Apart from performing more experiments to narrow the field, we prefer those differences that allow explanations that specify a mechanism, that are precise, and that contribute to the unification of our overall explanatory scheme. These preferences help us to move from the difference we consider to the causes we infer. But an account of this process of selection from the live candidates is only half of the story of the way we handle the problem of multiple differences. The other half is that we only consider a small portion of the actual and possible differences in the first place. We never begin with a full menu of all possible causal differences, because this menu would be too large to generate or handle. Yet the class of differences we do consider is not generated randomly. We must use some sort of short-list mechanism, where our background beliefs help us to generate a very limited list of plausible causal candidates, from which we then choose. While the full menu mechanism would have only a single filter, one that selects from all the possible differences, the actual short-list mechanism we employ has two stages, one where a limited list of live candidates is generated, the other where a selection is made from this list.

The need to use a short-list rather than a full menu raises a question about the scope of Inference to the Best Explanation. We

119

cannot say that the differences that do not make it onto the list are dismissed because they are judged to provide only inferior explanations. They are not dismissed on any grounds, because they are simply never considered. This raises a challenge for Inference to the Best Explanation. How can it account for the processes by which short-lists are generated? The principles of generation that solve much of the problem of multiple differences seem to depend on judgments of plausibility that do not rest on explanatory grounds: judgments of likeliness, but not of loveliness. A biological analogy may clarify what is at issue. The Darwinian mechanism of variation and selection is also a short-list mechanism. Natural selection does not operate on a full menu of possible variations, but only on the short list of actual variations that happen to occur. Here the process of generation is fundamentally different from the process of selection, so an account of the reasons only some types of individuals reproduce successfully would not seem able also to explain why only some types of individuals appear in a population in the first place. Similarly, the challenge to Inference to the Best Explanation is that the processes of generation and selection of hypotheses are fundamentally different, and that only the mechanism of selection depends on explanatory considerations.

I want to resist this restriction on the scope of Inference to the Best Explanation by suggesting how explanatory considerations can play a role in the generation of potential explanations as well as in the subsequent selection from among them. Let us extend the biological analogy. Darwin's mechanism faces the anomaly of the development of complex organs. The probability of a new complex organ, such as a wing, emerging all at once as a result of random mutation is vanishingly small. If only a part of the organ is generated, however, it will not perform its function, and so will not be retained. How then can a complex organ evolve? The solution is an appeal to 'preadaptation'. Complex organs arose from simpler structures, and these were retained because they performed a useful though perhaps different function. A wing could not have evolved all at once, and a half-wing would not enable the animal to fly, but it might have been retained because it enabled the animal to swim or crawl. It later mutated into a more complex structure with a new function. In one sense, then, mutations are not random. Some complex structures have a much higher probability of occurring than others, depending on whether they would build upon the simpler structures already present in the population. Mutations are not directed in the sense that they are likely to be beneficial but, since the complex organs that occur are determined both by the random process of genetic variation and the preadaptations already in place, only certain types of complex organs are likely to arise.

Preadaptations are themselves the result of natural selection, and they form an essential part of the mechanism by which complex organs are generated. So natural selection plays a role in both the generation and the selection of complex organs. Similarly, the mechanism of explanatory selection plays a role both in the generation of the short list of plausible causal candidates and in the selection from this list. The background beliefs that help to generate the list are themselves the result of explanatory inferences whose function it was to explain different evidence. We consider only the few potential explanations of what we observe that seem reasonably plausible, and the plausibility judgments may not seem to be based on explanatory considerations; but they are, if the background beliefs that generate them are so based. Those beliefs now serve as heuristics that guide us to new inferences, by restricting the range of actual candidates, much as preadaptations limit the candidate organisms that are generated (cf. Stein and Lipton, 1989). So Inference to the Best Explanation helps to account for the generation of live candidates because it helps to account for the earlier inferences that guide this process. The analogy to the evolution of complex organs also serves in another way. By exploiting preadaptations, the Darwinian mechanism yields the result that old and new variations fit together into a coherent, complex organ. Similarly, the mechanism of generating hypotheses favors those that are extensions of explanations already accepted, and so leads towards a unified general explanatory scheme. This scheme is itself a complex organ that could not have been generated in one go, but it is built on earlier inferences, themselves selected on explanatory grounds and guiding the generation of additional structures.

In the evolution of complex organs, the process of building on available preadaptations mimics what would be the result of a mechanism where natural selection operated on a full array of ready-made options, but only crudely. We find evidence of this difference in the imperfect efficiency of complex organs. We also

121

find much greater retention of old structures, traces of old fins in new wings, than we would if the complex organ had been selected from a comprehensive set of possible organs with a given function. Similarly, we should expect that the mechanism of considering only a short list of candidate explanations will generally yield different inferences than would have been made, had every possibility been considered before selecting the best. For example, we should expect to find more old beliefs retained under the short-list mechanism than there would be if we worked from a full menu of explanatory schemes, or from a random selection. Our method of generating candidate hypotheses is skewed so as to favor those that cohere with our background beliefs, and to disfavor those that, if accepted, would require us to reject much of the background. In this way, our background beliefs protect themselves, since they are more likely to be retained than they would be if we considered all the options. We simply tend not to consider hypotheses that would get them into trouble. The short-list mechanism thus gives one explanation for our apparent policy of inferential conservatism (cf. Quine and Ullian, 1978, pp. 66–8; Harman 1986, p. 46).

A really good version of Inference to the Best Explanation awaits a detailed account of explanatory loveliness. Fortunately, we do not have to wait for such an account before we compare and judge explanations, any more than we need to wait for a theory of induction before we make any inferences. And in the absence of a full account of what makes one explanation better than another, we still have reason to believe that explanatory considerations help us to solve the problem of multiple differences. Those considerations appear to play a role both in the process of generating hypotheses and in the process of selecting from among those that are generated. Since Inference to the Best Explanation can also account for this process as it applies to causes whose existence as well as causal role is inferred, this is enough to show that, as great as my debt to the Method of Difference has been, Inference to the Best Explanation is more than Mill.

VOLTAIRE'S OBJECTION

Let us now consider a different set of issues about the connection between explanatory loveliness and inference. In chapter four, I mentioned two general objections to the idea that explanatory considerations should be a guide to inference, that loveliness should be a guide to likeliness. One, 'Hungerford's objection' ('Beauty is in the eye of the beholder'), is that explanatory loveliness is too subjective and variable to give a suitably objective account of inference. The other, 'Voltaire's objection', is that Inference to the Best Explanation makes the successes of our inferential practices a miracle. We are to infer that the hypothesis which would, if true, provide the loveliest explanation of our evidence, is the explanation that is likeliest to be true. But why should we believe that we inhabit the loveliest of all possible worlds? If loveliness is subjective, it is no guide to inference; and even if it is objective, why should it line up with truth?

In reply to Hungerford's objection, the first thing to note is that inference is itself audience relative. Warranted inference depends on available evidence, and different people have different evidence: inferential variation comes from evidential variation. Moreover, as we have seen in the last section, inference also depends strongly on background beliefs, and these too will vary from person to person. It might, however, be argued that all background variation, and indeed all inductive variation, ultimately reduces to evidence variation, since a difference in beliefs, if rational, must reduce to a difference in prior evidence. But this is unlikely. It is not plausible to suppose that all scientific disagreements are due either to a difference in evidence or to irrationality. One reason for this is that, as Kuhn has emphasized, such disagreements sometimes stem from different judgments of the fruitfulness of scientific theory, of its ability to solve new puzzles or to resolve old anomalies. What counts is not just what the theory has explained, but what it promises to explain. This is clearly something over which rational investigators may differ, but the prospects for accounting for these differences entirely in terms of differences in available evidence are dim. Moreover, as I will argue in the next chapter, warranted inference may depend not only on the content of the evidence, but also on when it was acquired. Evidence that was available to the scientist when she constructed her theory may provide less support than it would have, had it been predicted instead. I will account for this difference by appeal to an inference to the best explanation that brings out the importance of distinguishing the objective support that a theory enjoys from scientists' fallible assessments of this support.

123

This distinction reveals another source of inferential variation that does not reduce to evidential variation.

So the simple claim that explanation is audience relative will not show that Inference to the Best Explanation is incorrect, since inference is audience relative as well. Hungerford's objection can also be defanged from the other side, by arguing that explanatory considerations do not have the strong form of relativity that the objection suggests. The explanatory factors I mentioned in the last section – unification, mechanism, precision, and so on – involve us in no more relativity on the explanatory side than they do on the inferential side, since they are the same factors in both cases. But the compatibility of Inference to the Best Explanation with a reasonable version of the interest relativity of explanation is perhaps clearest in the case of contrastive explanation. As I noted in chapter three, a contrastive analysis of why-questions illuminates the interest relativity of explanation by analyzing a difference in explanatory interest in terms of a difference in foil choice. Two people differ in what they will accept as a good explanation of the same fact, since one is interested in explaining that fact relative to one foil while the other has a different contrast in mind. Jones's syphilis will explain why he contracted paresis for someone who is interested in understanding why he, rather than Smith, who did not have syphilis, has paresis, but not for someone who wants to know why Jones contracted paresis, when other people with syphilis did not.

My account of contrastive explanation demystifies the phenomenon of interest relativity in two ways. First, by taking the phenomenon to be explained as a contrast rather than as the fact alone, it reduces the interest relativity of explanation to the truisms that different people are interested in explaining different phenomena and that a good explanation of one phenomenon will not in general be a good explanation of another. Second, as the difference condition on contrastive explanation shows, these differences in interest will require explanations that cite different but compatible elements of the causal history of the fact. This is no embarrassment for Inference to the Best Explanation, since that account allows us to infer many explanations of the same fact, so long as they are compatible. Moreover, the account allows that different people are sometimes warranted in inferring different contrastive explanations, since a difference in foil may correspond to a difference in experimental controls, and these differences clearly may be epistemically relevant. A difference in interest may correspond to a difference in evidence.

I conclude that Hungerford's objection has yet to be made in a form that threatens Inference to the Best Explanation. Explanation is audience relative, but so is inference, and we have been given no reason to suppose that the one relativity is more extreme than the other. Let us turn now to Voltaire. Why should the explanation that would provide the most understanding if it were true be the explanation that is most likely to be true? Why should we live in the loveliest of all possible worlds? Voltaire's objection is that, while loveliness may be as objective as you like, the coincidence of loveliness and likeliness is too good to be true. My first reply is that, as I hope to have convinced you over the past few chapters, it just turns out that our judgments of likeliness are guided by explanatory considerations. Perhaps this does make it surprising that our inductive methods should be successful, but the main purpose of the model of Inference to the Best Explanation has been to describe our practices, not to justify them. More, however, needs to be said, in part because Voltaire's objection might be taken to show that Inference to the Best Explanation is self-defeating. It might be claimed that the connection between loveliness and truth is so obscure that Inference to the Best Explanation could not provide as good an explanation of the success of the inferences it sanctions as some other account that showed that our inferences are really governed by a different mechanism.

One way to meet Voltaire's objection is to show that Inference to the Best Explanation does help to solve some of the problems of justification, that it helps to justify various aspects of our inferential procedures. I will consider this in some detail in the next two chapters. For now, we may observe that the structural similarity between simple contrastive inference and contrastive explanation that I have exploited in chapters three to six provides a kind of justification of some of our inferential practices, when they are construed as inferences to the best explanation. Insofar as one accepts that the Method of Difference is a reliable method for discovering causes, one ought also to accept that inferences to the best contrastive explanation are reliable, where those inferences coincide with the verdict of Mill's method.

There is, of course, a level of skepticism that would place the reliability of the Method of Difference in doubt as well, but at this

125

level Inference to the Best Explanation is in no worse shape than any other description of inference. At this level, the success of induction is miraculous or inexplicable on any account of the way it is done. This is one way of putting the conclusion of Hume's skeptical argument against induction. Hume himself used an over-simple 'More of the Same' description of inductive inference but, as I noted in chapter one, his skeptical argument does not depend on the particular description he gave. Whatever account one gives of our non-deductive inferences, there is no way to show *a priori* that they will be successful, because to say that they are non-deductive is just to say that there are possible worlds where they fail. Nor is there any way to show this *a posteriori* since, given only our evidence to this point and all *a priori* truths, the claim that our inferences will be successful is a claim that could only be the conclusion of a non-deductive argument and so would beg the question. In short, the impossibility of justifying induction does not depend on a particular account of our practices, but only on the fact that they are inductive. Consequently, in the absence of an answer to Hume, any account of induction makes inductive success miraculous, so the fact that Inference to the Best Explanation has this feature does not show that it is worse off than any other account. Inference to the Best Explanation would not have us infer that the loveliest possible world is actual; at most, it has us infer the loveliest of those worlds where our observations hold. And what Hume showed was that the success of any choice from among those worlds is inexplicable.

At this level of generality, there are only two positions that could claim an advantage over Inference to the Best Explanation. One is a deductivist account of scientific method, like Popper's, which takes Hume to have shown that we must abjure induction altogether. No such account, however, can be true to our actual inferential practices: we clearly do indulge in induction. The other is an account which grants that we use induction, but would have us use it more sparingly than Inference to the Best Explanation allows. A good example of this is Bas van Fraassen's 'constructive empiricism' (1980). In brief, van Fraassen's view is that we restrict inductive inferences to claims about observable phenomena. When a scientist accepts a claim about unobservable entities and processes, what she believes is only that it is 'empirically adequate', roughly, that its observable consequences are true. This contrasts with the realist version of Inference to the Best

Explanation that we have been considering in this book, since that account sanctions inferences to the truth (or the approximate truth) of the best explanation, whether it appeals to observables or not. So van Fraassen might claim that his account makes inductive success less miraculous than does Inference to the Best Explanation, for the simple reason that it requires fewer miracles.

I do not find constructive empiricism an attractive account. This is not the place for a detailed assessment, but I will mention two complaints. (I will have more to say about van Fraassen in chapter nine.) First, it is not clear that the account is consistent. Van Fraassen takes the view that what makes a claim observable is not that the claim employs only observation terms but that, however 'theory-laden' the description, what is described is something that our best theories about our sensory capacities tell us we can observe. To take one of his examples, we may describe a table as a swarm of electrons, protons, and neutrons without making the claim about the table non-observational (1980, p. 58). We cannot see a single particle, but we can see a swarm. Since the claim that my computer is now on a table is observable, and I am making the requisite observations, van Fraassen would have me believe that this claim is literally true. He would then also have me believe that 'my computer is on a swarm of particles' is literally true, not just empirically adequate. But I do not see how I can believe this true unless I believe that particles exist, which I take it is just the sort of thing I am not supposed to believe. (Am I supposed to believe that swarms of particles exist but individual particles do not?) My second complaint is that constructive empiricism gives a poor description of our actual practices, since we do actually infer the truth of claims about the unobservables. Most scientists do believe in electrons, protons, and neutrons and the claims they make about them, not just that these claims have true observable consequences; each of us believes that other people have had pains, itches, and visual impressions, though none of us can observe other people's phenomenal states. Moreover, the inferential path to unobservables is often the same as to unobserved observables. In these two sorts of case, the reasons for belief can be equally strong, so the suggestion that we infer truth in one case but not the other seems perverse. (I will return to this point in chapter nine.) Perhaps there could be creatures whose inductive mechanisms made them constructive empiricists, but they would be different from us.

Whatever one's view about the general merit of van Fraassen's position, however, it cannot claim an advantage over Inference to the Best Explanation with respect to Voltaire's objection, since the objection is not that it is inexplicable that explanatory considerations should lead to so many correct inferences, but that there should be any connection between explanatory and inferential considerations at all. (Nor do I think that van Fraassen claims such an advantage.) Voltaire's objection is not that Inference to the Best Explanation would make our inferences insufficiently parsimonious, but that it would make the success of the inferences we do make inexplicable. Moreover, as we will see in chapter nine, van Fraassen himself claims that there is such a connection, that the best explanation is one guide, not to truth in general, but to the truth of observable consequences (cf. 1980, pp. 23, 71). If someone claimed that patterns in tea leaves foretell the future, one might object that the connection is inexplicable, to which it would be no reply to say that the leaves are only reliable guides to the observable future.

Voltaire's objection is that the connection between explanation and inferential success is implausible because inexplicable, but Inference to the Best Explanation does not so far purport to justify our methods, only to describe them. Moreover, given Hume's skeptical argument, our successes seem inexplicable on any account. This is sufficient to answer Voltaire, but the objection suggests a related question that is worth considering. Given that our inferential methods are successful, why should they depend on explanatory considerations? Well, they have to depend on something, so why not on explanation? We have already dealt with the objection that explanatory considerations are too subjective to serve. Perhaps their use still seems unnatural as compared to 'More of the Same'. But a simple extrapolation model of induction will not cover the ground, since it will not account for vertical inference, so we need some less simple account. Second, Mill's methods provide a natural account of causal inference, though different from simple extrapolation and, as we have seen, the structure of contrastive explanations builds on this account. Lastly, Inference to the Best Explanation does rely on finding uniformities, both in the generalization from causal explanations of particular phenomena and in its general preference for unifying inferences. Inferences to the best explanation are not less natural than simple extrapolation.

There is something further that can be said to account for the match between loveliness and judgments of likeliness. Our judgments of likeliness change over time, in part because we gather new evidence, but our standards of explanation change too. Our explanatory standards, at one level of description, are malleable, so we can explain the match by saying that they have been molded to that purpose. Perhaps at one time astronomers held that the only satisfying explanation of the motion of the planets would have to show these to be the result of circular components, but later this standard was replaced by one that allowed for good explanation by appeal to elliptical motion and then by the more complex curves that result from the gravitational interaction of many bodies. Similarly, a Kuhnian account of exemplar-driven research suggests that exemplars mold a normal science tradition in part by setting a standard for good explanation, a standard that will change with a scientific revolution and the new exemplars it brings.

This, however, raises a priority question. Assuming that there is a match between loveliness and likeliness in part because our explanatory standards 'track' our judgment of likeliness, which is the horse and which is the cart? According to Inference to the Best Explanation, our explanatory standards guide our inferences, but why not say instead that we simply mold those standards to fit whatever we think likely, on other grounds? This would certainly take the bite out of Voltaire's objection: the reason the possible world we think likeliest to be actual is also the one with the loveliest explanations is simply that we adjust our explanatory standards to guarantee this result. But this view does not bode well for Inference to the Best Explanation. Explanation would not now be our guide to the truth, but only an epiphenomenon that makes us feel more at home in a world we have discovered by other means. Of course, on any realist account of explanation, actual explanation follows inference in the sense that we will only accept that we have given an actual explanation after we are convinced that the claims that are doing the explaining are correct, but to suppose that our view of what counts as a lovely potential explanation is determined by our inferences seems to rule out using explanatory considerations as a guide to inference.

By having explanation follow inference, we could explain the match between the two standards. We cannot give a parallel explanation if we have inference follow explanation, as Inference to the Best Explanation would have it, because our principles of inference, assuming they are sound, are answerable to an independent standard: the way the world is. Does this provide a strong argument against Inference to the Best Explanation? As you may by now imagine, I think not. For one thing, the epiphenomenalist will have to face a problem analogous to the one he presses against the explanationist. Since the epiphenomenalist does not think that explanatory considerations are our guide to inference, he will have to give some other account, and whatever factors an alternative account cites, the problem of explaining the match will reappear, for these factors and likeliness. Moreover, the burden of the last few chapters has been that Inference to the Best Explanation does a better job of accounting for our actual inferential practice than other accounts on offer. So this is an argument that explanation does act as a guide, and is not merely epiphenomenal. When we engage in contrastive inferences, we evaluate potential explanations of the contrast. On the epiphenomenal view, this interest in potential explanations is anomalous. We should only be concerned with actual explanations, or what we take to be actual explanations, and only after we have decided what is actual. From the point of view of Inference to the Best Explanation, however, this interest in potential explanations is just what we should expect, since it is by considering such explanations that we determine what is actual. Moreover, Inference to the Best Explanation helps to account for our particular explanatory interests, why we are more interested in explaining some things than others. According to it, we ought to have a particular interest in explaining the patterns, especially contrasts, in our evidence that are most revealing of the causal structure of the world, and this does seem to be where our interests lie.

There is another reason for placing explanatory considerations before inference. We are members of a species obsessed with making inferences and giving explanations. That we should devote so much of our cognitive energy to inference is no surprise, on evolutionary grounds. Knowledge is good for survival. That we should also be so concerned with understanding is more surprising, at least if explaining is just something we do after we have finished making our inferences. What good is this activity? If, however, Inference to the Best Explanation is along the right lines, then explanation has a central evolutionary point, since one of its functions is inference. It is, of course, possible that the practice of explanation has no evolutionary function: not all traits are adaptive. But the fact that Inference to the Best Explanation can account for the point of explanation in adaptive terms, while the epiphenomenal view cannot is, I think, an additional reason to favor it. This style of argument does not depend on the claim that we have a genetic predisposition to explain, though I find this likely, just as it seems likely that, whatever the actual nature of our inductive procedure, it has some innate basis. The same sort of consideration applies to our learned behaviors. Inference to the Best Explanation gives the practice of seeking and assessing explanations a central role in our own well-being, and so shows why we have all been so keen to learn how to do it.

The malleability of our explanatory standards in the history of science would explain the continuing match between judgments of likeliness and of loveliness but, if those standards are our guide to likeliness, what causes them to change? This is a difficult question, but the Kuhnian model of the dynamics of science may help to answer it. A tradition of normal science may enter a crisis state because the normal standards of explanation consistently fail to solve what that tradition identifies as some of its central problems or puzzles. This situation provides the motive for attempts to recast the tradition by revising those standards. If the new tradition succeeds in solving enough problems and in showing sufficient promise, this may convince scientists to make their inferences to the best explanation under the new standards, and we have a scientific revolution. On a conservative reading of this process, we may say that the *change* in explanatory standards is caused by a judgment of relative likeliness. The successes under the new standard are taken to be evidence that the inferences it sanctions are closer to the truth than those made under the old one. This, however, seems compatible with a strong version of Inference to the Best Explanation, since all the inferences within each tradition may still be guided by considerations of explanatory loveliness.

I do not claim that inference is the only point of explanation. This would probably not account for the extent of our interest in actual explanation. If the function of explanatory considerations were only to guide our inferences, then one might expect us to lose interest in explanations as soon as they have done their work and given us our inferences. In fact, however, there are good

131

inferential reasons to retain an interest in explanations after we have inferred that they are correct, since these explanatory beliefs will, in turn, serve as a guide to future inferences, as we saw in the last section. Because our system of beliefs evolves like a complex organ, we build new inferences on old explanations. Still, it seems unlikely that we only value actual explanations for this reason. If inference were the only point of explanation, this would probably not account for the strong causal asymmetry in explanation, for the fact that we allow that causes explain effects, but not conversely. (We do give teleological explanations, but I favor a causal account of these.) We often infer effects by first inferring causes, and Inference to the Best Explanation may allow for inference from as well as to the best explanation, as I suggested in chapter four, but this does not account for our refusal to count effects as explaining their causes. The asymmetry of explanation is not an objection to Inference to the Best Explanation, but it does suggest that inference is not the only point of explanation. Understanding is also an end in itself. This, however, does not undermine the claim that inference is one central function of explanation, or the argument that, by giving explanation this function, we have a more plausible account of our obsession with explanation than we would have, if we gave explanation no role in inference.

8

PREDICTION AND PREJUDICE

THE PUZZLE

The main purpose of the model of Inference to the Best Explanation is to provide a solution, or a partial solution, to the problem of description, by giving an illuminating account of the black box mechanism that governs our inductive practices. As I observed in chapter four, however, this model has also attracted philosophers because it describes a form of inference that appears to find useful application in philosophical argument, particularly in attempts to answer problems of justification, by showing that our inferential methods are reliable. In this chapter and the next, we will consider two such philosophical inferences to the best explanation. One is perhaps the best known argument of this form, the argument for scientific realism on the grounds that the approximate truth of predictively successful scientific theories is the best explanation of that predictive success. The other, the subject of this chapter, is the related but narrower argument that a theory deserves more inductive credit when data are predicted than when they are accommodated, on the grounds that only in the predictive case is the correctness of the theory the best explanation of the fit between theory and evidence.

In a case of *accommodation*, the scientist constructs a theory to fit the available evidence. If you are by now inclined to accept Inference to the Best Explanation as an account of scientific inference, read 'fit' as 'explain'. My discussion in this chapter, however, will not depend on this. If you are so recalcitrant as to continue to prefer some other account of the basic relation of inductive support, you may substitute that notion. For example, if you are a hypothetico-deductivist, a case of accommodation is

133

one where a theory is constructed so as to ensure that it, along with the normal complement of auxiliary statements, entails the evidence. In particular, my discussion about the putative advantage of prediction over accommodation will not rest on the claim that there is any deductive difference between the two cases, say that only predictions require deduction. To simplify the exposition, then, let us suppose that the explanations in question are deductive ones; the difference between explanatory and non-explanatory deductions will not affect the course of the argument. In a case of successful *prediction*, the theory is constructed and, with the help of auxiliaries, an observable claim is deduced but, unlike a case of accommodation, this takes place before there is any independent reason to believe the claim is true. (Because of this last clause, my notion of prediction is narrower than the ordinary one, closer perhaps to the idea of novel prediction.) The claim is then independently verified. Successful theories typically both accommodate and predict. Most people, however, are more impressed by predictions than by accommodations. When Mendeleyev produced a theory of the periodic table that accounted for all sixty known elements, the scientific community was only mildly impressed. When he went on to use his theory to predict the existence of two unknown elements that were then independently detected, the Royal Society awarded him its Davy Medal (Maher, 1988, pp. 274–5). Sixty accommodations paled next to two predictions.

Not all predictions provide as much inductive support or confirmation as Mendeleyev's, and some accommodations give stronger support than some predictions. Is there, however, any general if defeasible advantage of prediction over accommodation? We may usefully distinguish two versions of the claim that there is such an advantage. According to the 'weak advantage' thesis, predictions tend to provide more support than accommodations, because either the theory or the data tend to be different in cases of prediction than they are in cases of accommodation. For example, it might be that the predictions a scientist chooses to make are those that would, if true, provide particularly strong support for his theory. According to the 'strong advantage' thesis, by contrast, a successful prediction tends to provide more reason to believe a theory than the *same* datum would have provided for the *same* theory, if that datum had been accommodated instead. This is the idea that Mendeleyev's theory would have deserved less confidence had he instead accommodated all sixty-two elements. It seems to be widely believed, by scientists and ordinary people, if not by philosophers, that even the strong advantage thesis is correct. The purpose of this chapter is to determine whether this belief is rational.

However pronounced one's intuition that prediction has a special value over accommodation, it is surprisingly difficult to show how this is possible; so difficult that a number of philosophers have claimed that at least the strong thesis is false (e.g. Horwich, 1982, pp. 108–17; Schlesinger, 1987). The content of theory, auxiliary statements, background beliefs, and evidence, and the logical and explanatory relations among them, are all unaffected by the question of whether the evidence was accommodated or predicted, and these seem to be the only factors that can affect the degree to which a theory is supported by evidence. The difference between accommodation and prediction appears to be merely temporal or historical in a way that cannot affect inductive support. Moreover, the view that prediction is better than accommodation has the strange consequence that ignorance may be an advantage: the person who happens not yet to know about a certain datum would come to have more reason to believe his theory than someone who knew the datum all along. The problem is not merely that it seems impossible to analyze a preference for prediction in terms of more fundamental and uncontroversial principles of inductive support, but that the preference seems to conflict with those principles.

We can make the case against the strong thesis more vivid by considering a fictitious case of twin scientists. These twins independently and coincidentally construct the same theory. The only difference between them is that one twin accommodates a datum that the other predicts. If there really were an epistemic advantage to prediction, we ought to say that the predictor has more reason to believe the theory than the accommodator, though they share theory, data, and background beliefs. This is counterintuitive, but things get worse. Suppose that the twins meet and discover their situation. It seems clear that they should leave the meeting with a common degree of confidence in the theory they share. If they came to the meeting with different degrees of rational confidence in their theory, at least one of them ought to leave with a revised view. But what level should they settle on: the higher one of the predictor, the lower one of the accommodator, or somewhere in between? There seems no way to answer the question. Moreover, if there is a difference between prediction and accommodation, then the twin who should revise her view when she actually meets her sibling must not revise simply because she knows that someone like her twin might have existed. If revision were in order merely because of this possibility, then the difference between accommodation and prediction would vanish. Whenever data are accommodated, we know that there might have been someone who produced the theory earlier and predicted the data instead. But how can the question of whether there actually is such a person make any difference to our justified confidence in the theory? Any adequate defense of the putative difference between prediction and accommodation will have to explain how an actual meeting could be epistemically different from a hypothetical meeting. Those who reject the distinction seem on firm ground when they maintain that no such explanation is possible.

The case against the strong thesis seems compelling. Nevertheless, I will argue that the strong thesis is correct, that evidence usually provides more reason to believe a theory when it is predicted than it would if the evidence were accommodated. My argument will hinge on a distinction between the objective but imperfectly known credibility of a theory and the actual epistemic situation of the working scientist. Before I give my own argument, however, I will canvass some other plausible defenses of the strong thesis. We will find that none of them are acceptable as they stand but, in light of my own solution, we will be in a position to see the germs of truth they contain.

One of the attractive features of the puzzle of prediction and accommodation is that it requires no special philosophical training to appreciate. People are quick to catch on and to propose solutions. Some bite the bullet and deny that there is any difference. In my experience, however, most believe that prediction is better than accommodation, and give a number of different reasons. Three closely related ones are particularly popular. First, a theory built around the data is *ad hoc* and therefore only poorly supported by the evidence it accommodates. Second, only evidence that tests a theory can strongly support it, and accommodated data cannot test a theory, since a test is something that could be failed. Third, the truth of the theory may be the best explanation of its predictive success, but the best explanation of accommodation is instead that the theory was built to accommodate; since we accept the principle of Inference to the Best Explanation, then, we ought only be impressed by prediction.

None of these reasons for preferring prediction to accommodation are good reasons as they stand. Accommodating theories are obviously *ad hoc* in one sense, since *'ad hoc'* can just mean purpose-built, and that is just what accommodating theories are. To claim, however, that a theory that is *ad hoc* in this sense is therefore poorly supported begs the question. Alternatively, *'ad hoc'* can mean poorly supported, but this is no help, since the question is precisely why we should believe that accommodating theories are in this sense *ad hoc*. To assume that accommodating theories are *ad hoc* in the sense of poorly supported is to commit what might be called the *'post hoc ergo ad hoc'* fallacy. So the simple appeal to the notion of an *ad hoc* theory names the problem but does not solve it.

The second common reason given for preferring prediction is that only predictions can test a theory, since only predictions can fail. We are impressed by an archer who hits the bullseye, not by the archer who hits the side of a barn and then draws a bullseye around his arrow (Nozick, 1983, p. 109). But this argument confuses the theory and the theorist. A theory will not be refuted by evidence it accommodates, but that theory would have been refuted if the evidence had been different. Similarly, a theory will not be refuted by later evidence that it correctly predicts, though it would have been refuted if that evidence had been different. The real difference is that, although the predicting scientist might have produced his theory even if its prediction were to be false, the accommodating scientist would not have proposed *that* theory if the accommodated evidence had been different. It is only in the case of prediction that the scientist runs the risk of looking foolish. This, however, is a commentary on the scientist, not on the theory. In the case of archery, we want the bullseye drawn before the volley, but this is because we are testing the archer's skill. When we give students a test, we do not first distribute the answers, because this would not show what the students had learned. In science, however, we are supposed to be evaluating a theory, not the scientist who proposed it.

Finally we come to the appeal to an overarching inference to the best explanation. In the case of accommodation, there are two explanations for the fit between theory and data. One is that the theory is (approximately) true; another is that the theory was designed to fit the data. We know that the second, the accommodation explanation, is correct and this seems to pre-empt the inference to the truth explanation. In the case of prediction, by contrast, we know that the accommodation explanation is false, which leaves the truth explanation in the running. In one case, the truth explanation can not be the best explanation, while in the other it might be. This is why prediction is better than accommodation. This account has considerable intuitive force, but also a number of weaknesses. The most important of these is that it is unclear whether the accommodation explanation actually does pre-empt the truth explanation (Horwich, 1982, pp. 111–16). They certainly could both be correct, and to assume that accepting the accommodation explanation makes it less likely that the theory is true is once again to beg the question against accommodation. The issue is precisely whether the fact that a theory was designed to fit the data in any way weakens the inference from the fit to the correctness of the theory.

THE FUDGING EXPLANATION

I turn now to my own argument that prediction is better than accommodation. Notice first that, because of the 'short-list' mechanism we employ, which I discussed in the last chapter, there is some tendency for a theory that makes a successful prediction to be better supported overall than a theory generated to accommodate all the same data the first theory did, plus the datum the first theory predicted. The predicting theory must have had enough going for it to make it to the short list without the predicted datum, while the accommodating theory might not have made it without that datum. If this is so, then the predicting theory is better supported overall, even if we assume that the specific epistemic contribution of the datum is the same in both cases. But for just this reason, the short-list consideration does not itself argue for even the weak advantage thesis, since it does not show any difference in the support provided by prediction or accommodation itself, but at most only that the predicting theory will tend to have stronger support from other sources.

Let us consider now an argument for the weak advantage thesis. It is uncontroversial that some data support a theory more strongly than others. For example, heterogeneous evidence provides more support than the same amount of very similar evidence: variety is an epistemic virtue. Again, evidence that discriminates between competing theories is more valuable than evidence that is compatible with all of them. Similarly, the same set of data may support one theory compatible with the set more strongly than another since, for example, one theory may be simpler or in some other way provide a better explanation than the other. And there are many other factors that affect the amount of support evidence provides. Because of this, it is not difficult to find certain advantages in the process of prediction.

A scientist can choose which predictions to make and test in a way that she cannot choose the data she must accommodate. She has the freedom to choose predictions that will, if correct, provide particularly strong support for her theory, say because they will increase the variety of supporting evidence for the theory, or will both support the theory and disconfirm a competitor. This freedom is an advantage of prediction. Of course, just because scientists can choose severe tests does not mean that they will, but since scientists want to convince their peers, they have reason to prefer predictions that will, if correct, provide particularly strong support. Another advantage of prediction concerns experimental design. Much experimental work consists of using what are in effect Mill's methods to show that, when an effect occurs as a theory says it should, what the theory claims to be the cause of the effect is indeed the cause. The sort of controls this requires will depend on what cause the theory postulates. As a consequence, data that were gathered before the theory was proposed are less likely to have proper controls than data gathered in light of the theory's predictions, and so less likely to give strong support to the theory. Our discussion of Semmelweis in chapter five is a good example of this. Semmelweis was lucky enough to start with suggestive contrastive data, in the form of the difference in mortality due to childbed fever in the two maternity divisions of the hospital in which he worked. But he was able to get much better data after he proposed his various hypotheses, by designing experiments with careful controls that helped to discriminate between them. Theories thus provide guides to the sort of data that would best support them, and this accounts for a general advantage that predicted data tend to have over accommodated data.

So the weak advantage thesis is acceptable. Even if the actual support that a particular datum gives to a theory is unaffected by the time at which the datum is discovered, the data a scientist predicts tend to provide more support for her theory than the data she accommodates, because she can choose her predictions with an eye to strong support, and because she can subject her predictions to the sort of experimental controls that yield strong support. My main quarry in this essay, however, is the strong advantage thesis, and here considerations of choice and control do not help much. The freedom the scientist has to choose predictions that, if true, will provide strong support, does nothing to show that these predictions provide more support than the very same data would have, had they been accommodated. Similarly, the extra controls that the scientist can impose in cases of prediction should not be taken to underwrite the strong thesis, since an experiment without these controls is a different experiment, and so we ought to count the resulting data different as well. To defend the strong thesis, we need a new argument.

When data need to be accommodated, there is a motive to force a theory and auxiliaries to make the accommodation. The scientist knows the answer she must get, and she does whatever it takes to get it. The result may be an unnatural choice or modification of the theory and auxiliaries that results in a relatively poor explanation and so weak support, a choice she might not have made if she did not already know the answer she ought to get. In the case of prediction, by contrast, there is no motive for fudging, since the scientist does not know the right answer in advance. She will instead make her prediction on the basis of the most natural and most explanatory theory and auxiliaries she can produce. As a result, if the prediction turns out to have been correct, it provides stronger reason to believe the theory that generated it. So there is reason to suspect accommodations that does not apply to predictions, and this makes predictions better.

What I propose, then, is a 'fudging explanation' for the advantage of prediction over accommodation. It depends not so much on a special virtue of prediction as on a special liability of accommodation. Scientists are aware of the dangers of fudging and the weak support that results when it occurs. Consequently, when they have some reason to believe that fudging has occurred, they have some reason to believe that the support is weak. My claim is that they have such a reason when they know that the evidence was accommodated, a reason that does not apply in cases of prediction. It must be emphasized that this does not show that all accommodations have less probative force than any prediction. Similarly, it does not show that all accommodations are fudged. What it does show, or so I shall argue, is that the fact that a datum was accommodated counts against it in a way that would not apply if that same datum were predicted. The fudging explanation thus supports the strong advantage thesis.

An analogy may clarify the structure of the fudging explanation. Consider a crossword puzzle. Suppose that you are trying to find a word in a position where some of the intersecting words are already in place. There are two ways you can proceed. Having read the clue, you can look at the letters already in place and use them as a guide to the correct answer. Alternatively, you can think up an answer to the clue with the requisite number of letters, and only then check whether it is consistent with the intersecting letters. The first strategy corresponds to accommodation, the second to prediction. The first strategy is, I think, the more common one, especially with difficult puzzles or novice players. It is often only by looking at the letters that are already in place that you can generate any plausible answer at all. If, however, you are fortunate enough to come up with a word of the right length without using the intersections, and those letters are then found to match, one might hold that this matching provides more reason to believe that the word is correct than there would be if you had adopted the accommodating strategy for the word. If that is the case, why is it so? The only plausible explanation is that, when you accommodate, you have some reason to believe that the constraints that the intersections supply may pull you away from the best answer to the clue. Of course, assuming that the letters in place are correct, they do provide valuable information about the correct answer. But you have this information under both strategies, by the time you write in your word. It is only in the case of accommodation that the intersecting letters could possibly pull you away from the best answer to the clue, since it is only in this case that these letters are used in the process of generating the answer.

One might, of course, reject my suggestion that the two crossword strategies differ in the amount of support they provide. One might claim that there could be only a single word of the right length that both matches the intersections and fits the clue, or that one can simply see with either strategy how well a word fits the clue, so one need not take the difference in strategies into account. I do not think this is the case for all crossword puzzles, but nothing I will have to say about fudging in science depends upon this. The point of the crossword analogy is not to convince you that the fudging explanation is correct, but simply to illuminate its structure. As I hope will become clear, the case for saying that the motives for fudging support the strong thesis is much stronger in science than it is in crossword puzzles, because science offers so much more scope for fudging and because it is harder to detect. So let us return to science.

We can begin to flesh out the fudging explanation by distinguishing theory fudging and auxiliary fudging. Just as a theory may be more strongly supported by some data than by others, different theories may, as we have already noted, receive different degrees of support from the same data. This follows from the fact that there are always, in principle, many theories that fit the same data. If our inductive principles did not distinguish between these theories, the principles would never yield a determinate inference. So some theories must be more strongly confirmable than others, by a common pool of data that provides some support to all of them. When a theory is fudged, the result may be a theory that is only weakly confirmable. With special clauses and epicycles to handle particular accommodations, the theory becomes more like an arbitrary conjunction, less like a unified explanation. The result is that data that may support parts of the theory do not transfer support to other parts and the theory as a whole is only weakly supported. Another factor that affects the credibility of a theory is its relation to other non-observational claims the scientist already accepts. For example, as we discussed in the last chapter, a theory that is compatible with most of these background beliefs or, even better, coheres with them to produce a unified explanatory scheme is more credible than a theory that contradicts many of them or that depends on idiosyncratic causal mechanisms. The need for accommodation may force the scientist to construct a theory that fits poorly into the background and is thus harder to support. Prediction does not supply this motive to abjure the theory that fits best into the background.

142

When data that need to be accommodated threaten to force the scientist to construct a fudged theory, he can avoid this by tampering instead with the auxiliaries, but this too may result in weakened support. The theory itself may be elegant, highly confirmable in principle, and fit well into the background, but the accommodated data may only be loosely connected to it. This occurs since the scientist is forced to rely on auxiliaries that have little independent support, or approximations and idealizations that are chosen, not because they only ignore effects that are known to be relatively unimportant, but because they allow the scientist to get what he knows to be the right answer.

In short, in theory fudging the accommodated evidence is purchased at the cost of theoretical virtues; in auxiliary fudging it is at the cost of epistemic relevance. In both cases, the result is an inferior explanation. The fudging explanation suggests that the advantage of prediction over accommodation ought to be greatest where there is the greatest scope for fudging, of either sort, so we can provide some defense for the account by noting that this corresponds to our actual judgments. For example, we should expect the difference between prediction and accommodation to be greatest for complex and high-level theories that require an extensive set of auxiliaries, and to decrease or disappear for simple empirical generalizations, since the more auxiliaries, the more room for auxiliary fudging. I think that this is just what we do find. We are more impressed by the fact that the special theory of relativity was used to predict the shift in the perihelion of Mercury than we would have been if we knew that the theory was constructed in order to account for that effect. But when it comes to the low-level generalization that all sparrows employ a distinctive courtship song, we are largely indifferent to whether the observation that the song is employed by sparrows in some remote location was made before or after the generalization was proposed. Similarly, among high-level theories, we ought to judge there to be a greater difference between prediction and accommodation for theories that are vague or loosely formulated than we do for theories with a tight and simple mathematical structure, since vaguer theories provide more scope for theory fudging. Here again, this seems to be what we do judge. The fudging explanation also correctly suggests that the difference between accommodation and prediction ought to be roughly proportional to the number of possible explanations of the data we think there are. In a case where we convince ourselves that there is really only one possible explanation for the data that is, given our background beliefs, even remotely plausible (but not where we have simply only come up with one such explanation), fudging is not an issue and accommodation is no disadvantage. (I owe this point to Colin Howson.) And if the scientist is ever justifiably certain that she would have produced the same theoretical system even if she did not know about the evidence she accommodated, that evidence provides as much reason for belief as it would have, had it been predicted.

The fudging explanation gains some additional support from the way it handles two borderline cases. Consider first a case where a scientist has good independent reason to believe that some future event will occur, and ensures that the theory he constructs will entail it. It seems fair to say that those who admit a difference between prediction and accommodation would place this on a par with accommodation, even though it is a deduction of a future event. This is also entailed by the fudging explanation. The second case is the converse of this, an old datum, but one not known to the scientist when she constructs her theory and deduces that datum. This should be tantamount to a prediction, as the fudging explanation suggests.

The fudging explanation is related to Karl Popper's requirement that scientific theories must be incompatible with some possible observation (Popper, 1959). The falsifiability requirement does not directly discriminate between accommodation and prediction since, from a logical point of view, a theory that has only accommodated may be as falsifiable as one that has made successful predictions. This is why it was wrong to say simply that predictions are better than accommodations because only predictions are tests. Moreover, the difference in inductive support I am seeking between prediction and accommodation is precluded by Popper's deductivist philosophy, since he abjures the notion of inductive support altogether. Nevertheless, some sort of falsifiability is an important theoretical virtue, and there is sometimes the suspicion that a theory that only accommodates does not have it. The theoretical system may be so vague and elastic that it can be fudged to accommodate any observation. By contrast, a system that can be used to make a prediction will often be tight enough to be falsifiable in the broad sense that the theory, along with independently plausible auxiliaries, is incompatible

144

with possible observations. An astrologer who explains many past events does not impress us, but one who consistently made accurate predictions would give us pause. At the same time, the fudging explanation has considerably broader application than Popper's requirement. It also accounts for the difference between accommodation and prediction in the more interesting cases where the theories meet a reasonable version of the falsifiability requirement. Even if we are convinced that the accommodating theory is falsifiable, the suspicion remains that there was some fudging that weakened the data's inductive support, a suspicion that does not arise in the case of prediction. The use of unfalsifiable theories to provide accommodations is only the limiting case of a general phenomenon.

As I hope this discussion has made clear, what makes the difference between accommodation and prediction is not time, but knowledge. When the scientist doesn't know the right answer, she knows that she is not fudging her theoretical system to get it. The fact that her prediction is of a future event is only relevant insofar as this accounts for her ignorance. We can make essentially the same point in terms of the distinction between generating a theory and testing it. It follows from the definitions of accommodation and prediction that only accommodated data can influence the process of generation, and this is the difference that the fudging explanation exploits. What is perhaps somewhat less clear is that the fudging explanation is compatible with the claim that the objective degree of support that a theory enjoys is entirely independent of the time at which the datum is observed. We may suppose that this objective support the theory enjoys is precisely the same, whether the datum was accommodated or predicted. Nevertheless, the fudging explanation and the strong advantage thesis it underwrites may still be correct, because a scientist's actual epistemic situation gives her only fallible access to this objective support. As a consequence, information about whether a datum was predicted or accommodated is relevant to her judgment. But perhaps the best way to develop this important point and more generally to extend the case for the fudging explanation is to consider some objections.

First of all, it might be claimed that there is no real motive for fudging in the case of accommodation, since the scientist is free to choose any theoretical system that accommodates the available evidence. She can choose a natural and plausible combination of

145

theory and auxiliary statements that will therefore be highly confirmed by the accommodated data. The main weakness of this objection to the fudging explanation is that it severely exaggerates the ease with which a scientist can produce an accommodating system. Influenced by Quine, philosophers of science are eager to emphasize the underdetermination of theory by data, the fact that there are always in principle many theoretical systems that will entail any given set of data (Quine, 1951). This is an important point, but it may blind the philosopher to the actual situation of the working scientist, which is almost the opposite of what underdetermination suggests. Often, the scientist's problem is not to choose between many equally attractive theoretical systems, but to find even one. Where sensible accounts are scarce, there may be a great temptation to fudge what may be the only otherwise attractive account the scientist has been able to invent. The freedom to choose a completely different system is cold comfort if you can't think of one. And, in less dire straits, where there is more than one live option, all of them may need to be fudged to fit. This brings out an important point that the pejorative connotations of the word 'fudge' may obscure: a certain amount of fudging is not bad scientific practice. At a particular stage of research, it may be better to work with a slightly fudged theory than to waste one's time trying unsuccessfully to invent a pure alternative. Fudging need not be pernicious, and this is consistent with my argument that we ought to be more impressed by prediction than by accommodation.

Another line of thought that seems to undermine the fudging explanation begins with the idea that having extra evidence available when a theory is being constructed ought to make it easier to construct a good theory. But then it seems misguided to say that the fact that this evidence was available for accommodation somehow weakens the amount of support it provides. I accept the premise, with some reservations, but I reject the inference. Extra data can be a help, because they may rule out some wrong answers and suggest the right one. We are not, however, directly concerned with how easy it is to generate a theory, only with the reasons for believing a theory that has been generated, and we are not comparing the support a theory enjoys with this extra initial data to the support it enjoys without them. When a theory correctly predicts these data, it is not remotely plausible to claim that they would have provided more support if only the scientist had known about them when he constructed his theory. When the data are accommodated, however, this fact is reason to discount them (to some degree), since they provided a motive for fudging. This would not be plausible if evidence never misled us to an incorrect inference, but it often does.

The objections we have just considered were that neither prediction nor accommodation provide a motive for fudging. A third objection is that they both do. This has most force when focused on the auxiliary statements. For high-level theories, the route from theory to observation is long and often obscure. The only way to tie the theory deductively to the observational consequences will be with an elaborate and purpose-built set of auxiliary statements, including the battery of approximations, idealizations, and *ceteris paribus* clauses needed to make observational contact. This looks like fudging with a vengeance, and it seems to apply to prediction and accommodation alike. The natural response to this objection is to insist on distinguishing the manipulation that is necessary to get what one knows to be the right answer from the manipulation needed to generate any observational consequences at all. It is often difficult to get from a general and abstract theory to any empirical consequences, and so predictions may involve implausible auxiliaries that weaken the support the evidence eventually provides. In his ignorance, however, the predictor will try to use the best auxiliaries he can find, even if it turns out that this yields a false prediction. It is only the accommodator who will be tempted to use less plausible auxiliaries just to get the right result. So the difference between prediction and accommodation remains.

A fourth objection is that my explanation proves too much. As I observed at the beginning of this chapter, almost all theory construction involves some accommodation: a scientist rarely if ever constructs a theory without some data to which it is supposed to apply. This does not eliminate the question of the difference between prediction and accommodation, which is whether the fact that a particular datum was accommodated means there is any less reason to believe the theory than there would have been, had the datum been predicted. It might, however, seem to ruin the fudging explanation. If virtually every theory accommodates, then the fudging explanation seems to have the consequence that almost all predictive theories have been fudged, in which case there seems to be no general asymmetry between accommodation and prediction. My reply is twofold. First, while theory fudging may lead to a kind of global weakness in the theory that would make it implausible, poorly confirmable, and so only weakly supported by the evidence, this need not be the case with auxiliary fudging, since different auxiliaries may be used for different deductions. The fact that bad auxiliaries have been used to make an accommodation does not entail that they will be used to make a prediction. Indeed, the scientist will have a motive to use better auxiliaries when she turns to prediction. So the weak support provided by the accommodations is compatible with strong support from the successful predictions generated by the same theory.

My second and complementary reply brings out a new aspect of the fudging explanation. Accommodation is indirect evidence of fudging and, while this evidential link is quite general, it is also defeasible. I have already suggested that suspicions may be allayed in cases where the theory is low-level, requiring few auxiliaries, where it has a tight mathematical structure, or where it seems that there is no other possible explanation. Another way of allaying the suspicion of fudging is precisely by means of successful prediction. A fudged theory is one where the accommodation does not make it very likely that predictions will be successful; conversely, then, the success of predictions makes it less likely that the theory was fudged. Thus it turns out that the difference between accommodation and prediction according to the fudging explanation is not entirely a question of the potential liabilities of accommodation: prediction also has the special virtue that it may minimize them. This explains and defends what seems to be a common intuition that accommodations are most suspect when a theory has yet to make any tested predictions, but that the accommodations gain retrospective epistemic stature after predictive success. At the same time, we should not expect successful predictions to eliminate the relative weakness of accommodations entirely, in part because there may have been a switch in auxiliaries.

A fifth objection to the fudging explanation concedes the force of theory or auxiliary fudging in the case of accommodation, but maintains that there is a corresponding liability that applies only to prediction, namely 'observational fudging'. When a scientist makes a prediction, she has a motive to fudge her report of her future observation to fit it. There is no such motive in the case of accommodation, since the data are already to hand, and since in this case she can fiddle with the theoretical side to establish the fit between theory and evidence. According to this objection, accommodation and prediction both run risks of fudging, albeit of different sorts, so there is no clear advantage to prediction over accommodation. I agree that the risk of observational fudging is real. Moreover, it may be harder to detect than theory or auxiliary fudging, since most members of the scientific community can not check the investigator's raw data. (I owe this point to Morton Schapiro.) Observational fudging, however, applies to accommodation as well as to prediction. In accommodation there is sometimes, and perhaps usually, a mutual adjustment of theory and data to get a fit; in a successful prediction, only the data can be fudged, so the asymmetry remains. Of course, in cases of prediction where there is reason to believe that the observation has been badly fudged to score a success, we do not give the prediction much credit.

Scientists have ways of minimizing the risk of observational fudging. The most familiar of these is the technique of double-blind experiments. If a scientist wants to test the efficacy of a drug, he may ensure that he does not himself know which subjects are receiving the drug and which a placebo, so that he does not illicitly fulfill his own prophecy when he diagnoses the subjects' reactions. Inference to the Best Explanation gives a natural account of the value of double-blinds. When an experiment is done without one, the inference that would otherwise be judged the best explanation may be blocked, by the competing higher-level explanation that the fit between theory and reported data is due to unreliable reporting, skewed to make the theory look like the best explanation. Performing the experiment with a double-blind eliminates this competing explanation. The technique of double-blind experiments provides a useful analogy to the virtues of prediction as revealed by the fudging explanation. Just as the traditional double-blind is a technique for avoiding fudged observation, I am suggesting that the general technique of prediction provides an analogous technique for avoiding fudged theory. In one case, the double-blind improves the reliability of the observational reports; in the other, it improves the reliability of the theoretical system. Of course data gathered without the benefit of a double-blind may be unbiased, just as an accommodating system may not in fact include any fudging. In both cases, however, there is a risk of fudging, and the techniques of double-blinds and of prediction both reduce the risk that the process of inquiry will be polluted by wish-fulfillment.

ACTUAL AND ASSESSED SUPPORT

There is a final and particularly important objection I will consider. On my account, accommodation is suspect because it provides indirect evidence of fudging. It might be claimed that this evidence is relevant only if we know something about the theory's track record, but not much about the detailed content of the theory, auxiliaries, and data. The objection is that the information that the data were accommodated is irrelevant to the investigator himself. He has no need for indirect evidence of this sort, since he has the theory, data, and auxiliaries before him. He can simply *see* whether the theoretical system has been fudged and so how well his theory is confirmed by the evidence, whether that evidence is accommodated or predicted (Horwich, 1982, p. 117). In this case, to put weight on the indirect evidence provided by knowing the data were accommodated is like scrutinizing the tracks in the dirt when the pig itself is right in front of you (cf. Austin, 1962, p. 115). This objection does allow the fudging explanation some force. It admits that information about whether evidence was predicted or accommodated is relevant for someone who is unfamiliar with the details of the theory. It is also consistent with the claim that accommodating theories tend to be fudgier than predicting theories, hence that predictions are usually worth more than accommodations. It does not, however, allow that this information is relevant to the scientist who does the accommodating, or to any other scientist familiar with the details of the theoretical system and the evidence that is relevant to it.

According to this 'transparency objection', information about whether a datum was predicted or accommodated is irrelevant to a direct assessment of the credibility of a theory because, while credibility or support does depend on whether the theoretical system has been fudged, this is something that can be determined simply by inspecting the theory, the data, and their relation. We cannot, alas, simply observe whether a theory is true, but we can observe to what extent it is supported by the evidence. My reply is that even the investigator himself cannot merely observe the precise degree of support his theory enjoys. Inductive support is translucent, not transparent, so the indirect evidence that the fudging explanation isolates is relevant to both scientist and spectator. What I am suggesting is that we need to distinguish between actual and assessed inductive support, between the extent to which the data actually render the theory probable and the scientist's judgment of this. As I formulated it at the beginning of this chapter, the strong advantage thesis claims that a predicted datum tends to give more reason to believe a theory than the same datum would have provided for the same theory, if that datum had been accommodated. It does not claim that the actual support for the theory is different, only that our assessment of this support ought sometimes to be different. Let us see how this can be so.

It ought to be uncontroversial that there is a distinction between actual and assessed support, since it is a familiar fact of ordinary as well as scientific life that people misjudge how well their views are supported by their evidence. Moreover, without this distinction, it is difficult to see how we could account for the extent of scientific disagreements. We should expect support to be only translucent, even with respect to those factors that depend exclusively on the content of the theory, auxiliaries, and evidence. A scientist may misjudge the variety of the evidence or the simplicity of his theory. He may also be wrong about the plausibility of the auxiliaries, since these are rarely all made explicit. But the case for saying that scientists are only fallible judges of the actual support their theories enjoy is even stronger than this, since actual support also depends on additional factors that go beyond the content of the theoretical system and the evidence. First, it depends on the relationship between that system and the scientist's complex web of background beliefs. Second, it depends on the existence of plausible competing theories, theories the scientist may not know. Lastly, as Kuhn has suggested, it depends on the fruitfulness of the theory, on its promise to solve new puzzles or old anomalies. This may be an objective matter, and it is a consideration in judging the credibility of a theory, but it is certainly not 'given'. For all these reasons, it seems clear that support is translucent, not transparent. Scientific claims are all fallible, whether they concern theory, observation, or the bearing of one on the other.

151

Since support is translucent, the fudging explanation applies to the judgments of the scientist herself. The actual support that evidence gives to theory does not depend on the information that the evidence was accommodated or predicted but, since we can only have imperfect knowledge of what this support is, the information is epistemically relevant. It is worth scrutinizing the tracks in the dirt if you cannot be certain that what you see is a pig. Fudging need not be a conscious process, so the scientist should not assume that she is not doing it just because she is not aware that she is. The mechanism by which scientists judge credibility is not conscious either, as I emphasized in chapter one, or else the problem of describing their inductive practices would not be as extraordinarily difficult as we have found it to be. So there is plenty of room for unconscious and undetected fudging. This is why the indirect evidence that information about whether data were accommodated or predicted provides remains relevant to scientists' assessment of support.

The failure to acknowledge the gap between actual and assessed credibility has a number of probable sources. One is the legacy of a too simple hypothetico-deductive model of induction, since deductive connections are meant to be the model of clarity and distinctness. But once we appreciate how complex judgments of credibility are, we should be happy to acknowledge that these are fallible. If the fudging explanation is along the right lines, we should also conclude that judgments of credibility have an empirical component, since the question of whether evidence has been accommodated is an empirical one. Another reason the distinction between actual and assessed support has been elided, or its importance missed, is the tradition of stipulating an artificial division between the context of discovery and the context of justification, between the ways theories are generated and the ways they are evaluated. The difference between accommodation and prediction is a difference in generation, but what the fudging explanation shows is that this is relevant to the question of evaluation.

The assumption that support is transparent to the investigator, that no distinction need be drawn between actual and assessed support, is an idealization, very similar to the idealization epistemologists sometimes make that people are deductively omniscient, so that they know all the deductive consequences of their beliefs. Such an idealization may be useful for certain issues, but it obscures others. It would, for example, yield an absurd account of the nature of philosophy or of mathematics. Again, to give a scientific analogy, the assumption of transparency is like a biological idealization that all members of a species are identical. This is a simplification that might serve various biological purposes, but it would obscure the mechanism of natural selection, by ignoring what makes it possible. Similarly, the idealization of transparency, though perhaps sometimes useful, obscures the mechanism that explains why prediction is better than accommodation, and the reason this information is relevant to scientists' assessments of the merits of their theories.

Only accommodations influence the generation of theory, and theory influences only the generation of predictions. These two observations have provided the basis for my argument for the epistemic advantages of prediction over accommodation. Predictions are better, because the theory under investigation may be used to select those that would give particularly strong support, especially because they would discriminate between this theory and its rivals. Accommodations are worse, because they may lead to a theory that is only weakly supportable. In accommodations, the scientist knows the correct answers in advance, and this creates the risk, not present in predictions, that he will fudge his theory or his auxiliaries. Moreover, since a scientist's judgment of the degree of support evidence provides is only a fallible assessment of its actual value, the indirect evidence of fudging that the fact of accommodation provides is relevant to the scientist's own judgment of support. It leaves the scientist with less reason to believe his theory than he would have had, if he had predicted the datum instead.

One clear limitation of my account is that it does not include anything like a precise characterization of the features that make one theoretical system fudgier than another. I have not provided such a characterization because I do not know how to do so, but my argument does not depend on it. It is enough that we have good reason to believe that different theoretical systems enjoy different degrees of support from data they all fit, and that the requirements of accommodation may be in tension with the desire to produce a highly confirmable theoretical system. Another limitation is that I have not shown that the fudging explanation accounts for the true extent of the contrast between prediction and accommodation. Perhaps most of us feel a greater contrast than my account appears to justify. There are then various possibilities. One might simply say that this excess is psychologically real, but rationally unjustifiable. There is some advantage to the fact of prediction, but it is weaker than many suppose. Second, one might argue that my account justifies a greater contrast than it initially seems to do. Two of the ways this might be done are by stressing the way the case for the weak advantage thesis combines with the case for the strong thesis, or by emphasizing the retrospective credit that successful predictions grant to earlier accommodations. Finally, one might argue that there is some independent and additional reason why predictions are better than accommodations that I have not taken into account. My story does not rule out such a thing but, given the difficulty in coming up with any justification at all for the strong advantage thesis, I am content to have given one.

In the first section of this chapter, I criticized several plausible accounts of the difference between accommodation and prediction. From the perspective of the fudging explanation, we are now in a position to see the germs of truth they contain. Recall first the claim that accommodations are worse than predictions because accommodations are *ad hoc*. My objection was that this leaves the question unanswered or begged, either because calling a theoretical system *ad hoc* simply means it is designed to accommodate, or because it means it is only weakly confirmable. The fudging explanation provides an independent reason why accommodating systems tend to be *ad hoc* in the second sense, only weakly supported by the evidence they accommodate.

Another account I initially rejected was that predictions are better because only they test the theory, since a test is something that can be failed. My objection was that, while it may be that only predictions test the scientist, an accommodating theory can be as falsifiable as a predicting one. In such a case, if the accommodated evidence had been different, the theory would have been disconfirmed. As we have seen, however, in extreme cases there will be a suspicion that an accommodating theory is in fact unfalsifiable, a suspicion that does not arise in the case of prediction. More generally, the fudging explanation is related to the notion that we should test the scientist, and that she should test herself. It is not simply that she ought to run the risk of being wrong, though that is a consequence of the real point. She should place herself in a situation where, as in the case of a standard double-blind experiment, she does not know the right answer in advance.

At long last, we return to the claim that an overarching inference to the best explanation underwrites the strong advantage thesis. In its standard form, that account claims that prediction is better because the best explanation for predictive success is truth, while the best explanation for accommodation is instead that the theory was designed for that purpose. My objection was that it was not clear that the accommodating explanation pre-empts the truth explanation, making it less likely. It is true that only in cases of accommodation can we explain the fit between theory and data by pointing out that the theory was designed for the purpose, but whether this makes it less likely that the theory is correct is just what is at issue. By contrast, it is clear that the fudging explanation competes with the truth explanation. Insofar as we may reasonably infer the explanation that the fit between theory and data in the case of accommodation is due to fudging, this undermines our confidence in the inference from fit to truth. So the best explanation account of the difference between accommodation and prediction can be salvaged by replacing the accommodation explanation with the fudging explanation. In cases of accommodation, the inference from fit to truth may be blocked by the inference from fit to the conclusion that the theoretical system is *ad hoc*, in the epistemically pejorative sense. In addition, we may now say that we have vindicated the original claim that the accommodation explanation and the truth explanation are competitors. We have shown this in two steps, by arguing that accommodation is evidence of fudging, and that fudging results in weakened support. For reasons that will become clear in the next chapter, however, it is perhaps better to say simply that the fudging explanation competes with the ordinary inferences to the best explanation that scientists are otherwise inclined to make, rather than speaking of a special, philosophical 'truth explanation'. Seen in this way, the fudging explanation should sometimes act as a brake on inference, preventing scientists from accepting what they otherwise would judge to be an explanation lovely enough to infer.

This view of the way an overarching inference to the best explanation accounts for the epistemic distinction between prediction and accommodation links up with several features of Inference to the Best Explanation that I have discussed earlier in this book, of which I will mention only two. The first is that it allows us to give a more nuanced response to Hungerford's objection (chapter seven). That objection is that explanatory loveliness is audience relative in a way that makes it unsuitable as an objective notion of inductive warrant. My main response was that, as the structure of contrastive explanation shows, loveliness is relative, but in an innocent way, since different interests will lead to different but compatible inferences. Armed now, however, with the distinction between actual and assessed support, we may add that while actual loveliness is not strongly relative, assessed support may be. Different scientists may judge incompatible theories loveliest, if at least one of them is wrong. But scientists also have ways of minimizing the dangers of this sort of misassessment, and the special value they place on novel prediction over accommodation is one of them. The fudging explanation also brings out one consequence of the short-list method that we must employ in inference (chapter seven). If Inference to the Best Explanation worked from a full menu of all possible explanations of the evidence, the fudging explanation would not apply. The scientist would have no motive to fudge her explanation to make the accommodation, since she would have before her every possible explanation, from which she could choose the best. Since, however, this can never be her position, and since she often must scramble to generate even one plausible candidate, the fudging explanation applies. Thus the epistemic distinction between accommodation and prediction is one symptom of the way our actual inferential techniques only imperfectly mimic a full menu mechanism.

I end this chapter by observing that the fudging explanation also gives us an answer to the puzzle of the twin scientists, which seemed to show that there could be no difference between prediction and accommodation. We had two scientists who happen to come up with the same theory, where one accommodates evidence the other predicts. After they compare notes, they must have a common level of confidence in the theory they share. The difficulty was that it seemed impossible to say what level this should be, and how meeting a predictor could be any different for the accommodator than knowing what all accommodators know, namely that if someone had produced the same theory on less evidence, her prediction of the balance of the evidence would have been successful. The answer to this puzzle is now clear. The accommodator ought to worry that she had to do some fudging, but her suspicion is defeasible. One of the things that can defeat it, though not common in the history of science, is meeting a corresponding predictor. If the accommodator meets someone who predicted data she only accommodated, with the same theoretical system, this shows that she almost certainly did not fudge to make those accommodations. The predictor, ignorant as he was of the data he predicted, had no motive to fudge his theoretical system to get those results; consequently, the fact that he came up with just the same system provides strong independent evidence that the accommodator did not fudge either. At the same time, the fact that any accommodator knows that the same theory could have been constructed earlier is not enough, since such a construction might have then required arbitrary and unmotivated fudging. A predictor might have come up instead with a different and better theoretical system. If an actual meeting takes place, however, the twins should leave it sharing the higher confidence of the predictor. The accommodator, like all scientists, has only fallible knowledge of the actual credibility her theory enjoys, and the meeting gives her some additional evidence that ought generally to lead her to increase her assessment.

9

TRUTH AND EXPLANATION

CIRCULARITY

We have just seen one way Inference to the Best Explanation can be used to provide a justification for a particular aspect of our inductive practices. We tend to give more credit to successful predictions than to accommodations, and the fudging explanation shows how this preference can be justified, by showing how it follows from a more general preference for the best explanation. Accommodations are often worth less than predictions, because only they have to face the possibility that the best explanation of the fit between theory and data is that the theoretical system was fudged. In this chapter, we will consider a second overarching application of Inference to the Best Explanation to a problem of justification, where now what is to be justified is not just one rather special aspect of our inductive practices, but the general claim that they take us towards the truth.

When a scientist makes an inference to the best explanation of the sort we have discussed in past chapters, I have taken it that she infers that the claim in question is (at least approximately) true, whether the claim concerns something observed, something unobserved but observable, or something unobservable. Even if this descriptive claim is accepted, however, it leaves open the question of justification. Do we have any reason to believe that the inferences to the best explanation that scientists make really are truth-tropic, that they actually take scientists towards the truth? To have a familiar label, let us call someone who accepts both that Inference to the Best Explanation provides a good description, and that this form of inference is truth-tropic, a 'scientific realist'. There is a well-known argument for scientific realism that itself has the form of an inference to the best explanation. In its simplest version, the argument is that we ought to infer that scientific theories that are predictively successful are (approximately) true, since their truth would be the best explanation of their success (cf. Putnam, 1975, p. 73, n. 29; 1978, pp. 18–22). Moreover, since these theories are themselves the results of inferences to the best explanation, that form of inference is truth-tropic. Let us call this argument from predictive success to truth tropism the 'truth argument'.

A realist who endorses the truth argument will take it that Inference to the Best Explanation is a generally legitimate form of inductive inference, and so will endorse the inferences of that form that scientists make. The truth argument itself, however, is not supposed to be a scientific inference; instead it is a philosopher's argument, an additional inference of the same form, whose conclusion is that the scientific inferences are truth-tropic. What is the intuition behind this argument? Suppose that you find yourself in the middle of the woods with a detailed topographical map that may or may not be accurate. You then proceed to navigate by means of the map, and find that everything the map says you ought to find – rivers, lakes, roads, peaks, and so on – is just what you do find, though of course you only see a small portion of the terrain the map purportedly depicts. What is the best explanation of the success you have achieved with this map? One possibility is that your success is a fluke: the map happens to be correct in those few details you have checked, but it is generally wrong. In this case, however, your success is inexplicable: it is simply good luck. Another possibility, however, is that the map is generally accurate, both in the parts you have checked and in those you have not. Its general accuracy would explain why you have had success using it, and this is why you would infer that the map is accurate. Similarly, the predictive success of a scientific theory does not entail that the theory is correct, since every false statement has many true consequences. So one 'explanation' for predictive success is that the predictions that have been checked just happen to be some of the true consequences of a false theory. This, however, is really no explanation at all, since it is just to say that the predictive success is a fluke. By contrast, if the theory is true, this would explain its predictive success. So, since Inference to the Best Explanation is a warranted form of inference, we have reason to infer that the theory is true.

This is an inductive argument, so it does not prove that successful theories are true, but it does, according to its proponents, provide a good reason for believing that they are true, and so that the form of inference that led to them, namely Inference to the Best Explanation, is a reliable guide to the truth.

I want to consider whether the truth argument is cogent, taking seriously its pretension to be both distinct from the particular inferences scientists make and a real inference to the best explanation, as that form of inference has been developed and defended in this book. The most obvious objection to the truth argument is that it begs the question (cf. Laudan, 1984, pp. 242–3; Fine, 1984, pp. 85–6). Who is the argument supposed to convince? It is an inference to the best explanation, so it has no force for someone who rejects inferences to the best explanation, either because, like Popper, she rejects inductive inferences altogether or because, while she accepts some form of induction, she does not accept inferences to the best explanation. More surprisingly, it does not even have force for everyone who accepts some version of Inference to the Best Explanation. Recall, from chapter four, that Inference to the Best Explanation requires that we distinguish between potential and actual explanations. We cannot say that we are to infer that the best actual explanation is true, since to say that something is an actual explanation is already to say that it is true, so this method would not be effective. Instead, we must say that we are to infer that the best potential explanation is true, or that it is an actual explanation. Because of this feature of Inference to the Best Explanation, the model may be co-opted by someone who denies that inferences generally are or ought to be inferences to the truth of the inferred claim. One might claim that the best potential explanation is a guide to inference, yet that what we ought to infer is not that the explanation is true, but only, for example, that its observable consequences are true. As I mentioned in chapter seven, Bas van Fraassen holds this view, in some of his writings. Such a person would not be moved by the truth argument, at least not in the direction its proponents intend. Perhaps the truth of a theory is the best explanation of its predictive success but, if the theory traffics in unobservables, the claim that it is true is not an observable claim, so the most that a constructive empiricist like van Fraassen will accept is that the theory has only true observational consequences. The realist is trying to argue from a theory's past observational successes to its truth, not from past observational success to future ones. (Van Fraassen himself rejects the truth argument on the different ground that truth is not a good explanation of success. We will consider this objection below.) So there is at most one sort of person who ought to be impressed by the truth argument, and this is someone who both accepts Inference to the Best Explanation and accepts that this form of inference is truth-tropic. In other words, the truth argument can only have force for a scientific realist. This, however, is precisely the view that the truth argument was supposed to defend. In short, the truth argument is an attempt to show that Inference to the Best Explanation is truth-tropic by presupposing that Inference to the Best Explanation is truth-tropic, so it begs the question.

The circularity objection shows that the truth argument has no force for the non-realist. The interesting remaining question is whether it may yet be a legitimate argument for someone who is already a scientific realist, giving her some additional reason for her position. To answer this, it is useful to compare the truth argument to the general problem of justifying induction. The truth argument is an attempt to justify a form of inductive inference, and so we should have expected it to run up against the wall that Hume's skeptical argument erects against any such attempt. More particularly, the circularity problem for the truth argument is strikingly similar to the problem of trying to give an inductive justification of induction. Most of us have the gut feeling that the past success of induction gives some reason for believing that it will continue to work well in the future. This argument, however, is itself inductive, so it cannot provide a reason for trusting induction. The counter-inductive justification of counter-induction, which claims that counter-induction will work in the future because it has failed in the past, strikes us as obviously worthless, but the only relevant difference between this and the inductive justification of induction seems to be that the latter confirms our prejudices (Skyrms, 1986, sec. II.3). The truth argument has just the same liability as the inductive justification of induction: it employs a form of inference whose reliability it is supposed to underwrite, and this seems to show not only that the argument is of no value against the skeptic, but that it has no value for anyone.

Objections from circularity are among the most elegant and effective tools in the philosopher's kit, but I am not sure whether the circularity objection is conclusive against either the inductive justification of induction or the truth argument. One reason I hesitate is that, for all its intuitive appeal, circularity is very difficult to characterize. We cannot appeal simply to the psychological force of an argument, since people often reject perfectly good arguments and are often persuaded by circular arguments. Many are convinced by Descartes' argument that every event must have a cause, since a cause must have at least as much 'reality' as its effect, and this would not be so if something came from nothing. Yet, as Hume observed, the argument is circular, since it implicitly treats 'nothing' as if it would be a cause, which would only be legitimate if it were already assumed that everything must have a cause. Nor will it do to say that an argument is circular just in case its conclusion is somehow already present among its premises. This runs the risk of counting all deductive arguments as circular. In fact, a deductive argument may be legitimate even though its conclusion is logically equivalent to the conjunction of the premises. The classic free will dilemma is an example of this. Either determinism is true or it is not; if it is, there is no free will; if it is not, there is no free will; therefore there is no free will. The premises clearly entail the conclusion and, on a truth functional interpretation of the conditionals, which seems not to diminish the force of the argument, the conclusion entails each of the premises, so the conclusion is logically equivalent to the conjunction of the premises. Yet this is, I think, one of the stronger arguments in the history of philosophy, even though its conclusion is incredible. A better account of circularity is possible if we help ourselves to the notion of a good reason. We cannot say that an argument is circular if there is less reason to accept the premises and rules of inference than there is to accept the conclusion, since this condition is met by the non-circular argument that the next raven I see will be black because all ravens are black. But we might say that an argument is circular just in case the conclusion is an essential part of any good reason we might have to accept one of the premises or rules of inference. I suspect, however, that any account of circularity that relies on the notion of good reason is itself circular, since we cannot understand what it is to have a good reason unless we already know how to distinguish circular from non-circular arguments.

Another reason I hesitate to dismiss the inductive justification of induction is that there are clear cases where we can use inductive arguments to defend a form of inductive inference. I will give two examples from personal experience. The first is the charting method. Many years ago, I spent a summer working as a clerk on the London Metal Exchange. Although my responsibilities were menial, I took an interest in the way dealers formed their views on the future movements of the markets in the various metals they traded. The dealers I spoke to cited two different methods. The first, the physical method, is straight-forward. It consists of monitoring those conditions which obviously affect the supply and demand of the metal in question. For example, a strike at the large Rio Tinto copper mine was taken as evidence that copper prices would rise. The second, the charting method, is more exotic. Chartists construct a graph of the prices of the metal over the previous few months and use various strange rules for projecting the curve into the future. For example, two large hills (whose peaks represent high prices) separated by a narrow valley would be taken as an indication that prices would rise, while a small hill followed by a larger hill suggested that prices would fall. The dealers provided no explanation for the success of this method, but many of them put stock in it. My second example is persistence forecasting. During the summer after the one on the Metal Exchange, I spent several weeks hiking and climbing in the Wind River mountain range of Wyoming. In these travels I met another hiker who turned out to be a meteorologist. After some talk about the weather, our con-versation turned to predicting it. She told me that there had been studies comparing the reliability of various predictive methods and that the technique of persistence forecasting still compared favorably with other methods. Persistence forecasting, it turns out, is the technique of predicting that the weather will be the same tomorrow as it was today.

Perhaps the charting method should be of more interest to the anthropologist than to the epistemologist, but it does provide a clear case of an inductive method whose reliability could be assessed with an inductive argument based on past performance. The method is implausible, but an impressive track record would lead us to give it some credit. Similarly, though I do not know whether the meteorologist was being entirely honest about its performance, persistence forecasting is a technique whose reliability can be sensibly assessed by considering how well it would have worked in the past. This example is particularly attractive to the defender of the inductive justification of induction, since, unlike charting, persistence forecasting is itself a kind of primitive enumerative induction. The general point of both examples, however, is to show that we can legitimately assess inductive methods inductively.

If charting and persistence forecasting had strong track records, this would not impress an inductive skeptic, but it would impress us, since we already accept induction. Similarly, the fudging explanation as an argument that predictions are better than accommodations might not move someone who rejected Inference to the Best Explanation altogether, but it still has some force for the rest of us. What this suggests is that the notion of circularity is audience relative. The same argument may beg for one audience, yet be cogent for another. So while the inductive justification of induction has no force for an inductive skeptic, it may yet have some value for us (cf. Black, 1970). There is nothing illegitimate about giving arguments for beliefs one already holds. Yet we are pulled by conflicting intuitions in this case. On the one hand, the Humean argument for the circularity of the inductive justification does not remove the strong feeling that the past successes of induction bode well for its future performance. On the other hand, we also feel that the pathological ineffectiveness of the inductive justification of induction against someone who does not already accept induction shows that it also provides no good reason for those of us who do accept it. This is why being circular may be more debilitating than merely having premises your opponent does not accept. Someone who is inclined to doubt my veracity will not be convinced of something on my say-so, and he should not be moved by my additional statement that my testimony is true. Someone else may have reason to trust me, but my additional statement provides him with no additional reason. In short, the debilitating consequences of circularity do not always disappear when an argument is used to preach to the converted.

The fact that arguments from the track records of charting and persistence forecasting would beg against an inductive skeptic does not show that these arguments are circular for us, yet the fact that the track record of induction would beg against the skeptic seems to show that it also begs for us. What is the difference? In the former cases, it is easy to see how there could be people who initially reject charting and persistence forecasting, yet accept induction. So there are people who start by having no reason to trust these specific methods, yet ought, by their own principles, to accept the arguments from track record. In the case of the inductive justification of induction, by contrast, it is hard to see how the argument could have any force for anyone who did not already accept its conclusion, since such a person could only be an inductive skeptic. The underlying idea here is that, unless an argument could be used to settle some dispute, it can have no force even for people who already accept its conclusion.

But the track record of induction could be used to settle a dispute, not over whether induction is better than guesswork, but over the *degree* of reliability of induction among disputants who endorse the same principles of induction. These people would give different weight to the inductive justification, but they would all give it some weight. Consider someone who is much too optimistic about his inductive powers, supposing that they will almost always lead him to true predictions. Sober reflection on his past performance ought to convince him to revise his views. If he admitted that he had not done well in the past, yet claimed for no special reason that he will be virtually infallible in the future, he would be inductively incoherent. Similarly, someone who is excessively modest about his inductive powers, though he gives them some credit, ought to improve his assessment when he is shown how successful he has been in the past. And the fact that the inductive justification can help to settle disputes over reliability is enough, I think, to show that even people who already believe that induction has the degree of reliability that the argument from its track record would show, can take that argument to provide an additional reason for holding that their belief is correct. So the inductive justification of induction, while impotent against the skeptic, is legitimate for those who already rely on induction. If the problem of induction is to show why the skeptic is wrong, the inductive justification of induction is no solution, but it does not follow that the justification has no application. Someone who has no confidence in my testimony will not be moved if I add that I am trustworthy, but someone who already has some reason to trust me will have an additional reason to accept my claim if I go on to say that I am certain about it.

There is an additional reason for saying that the inductive justification is legitimate for those who already accept induction.

Consider again charting and persistence forecasting, but imagine now that we have two machines that generate predictions, one about metal prices, the other about the weather. In fact, one machine runs a charting program, the other a persistence program, but we do not know this. All we notice is that these machines generate predictions, and we then might use our normal inductive techniques to evaluate their reliability, much as we use those techniques to determine how reliable a drop in barometric pressure is as a predictor of tomorrow's weather. Well, some organisms are predicting machines too. Suppose we find alien creatures who predict like mad, though we do not yet know what mechanism they use. Still, we may look at how well these creatures have done in the past to assess their future reliability. Our procedure will not be undermined when we later discover their predictive mechanism is the same as ours, though this will show that we have unwittingly given an inductive justification of induction.

Our problem is to determine when an argument that is circular for some may nevertheless provide a reason for belief among those who already accept its conclusion. My suggestion is that this form of preaching to the converted is legitimate when there might also be someone who does not already accept the conclusion of the argument but who would accept the premises and the rules of inference. In short, an argument provides an additional reason for those who already accept its conclusion, when there is a possible audience which does not already accept it, yet for which the argument would not be circular. This condition does not give us an analysis of circularity, but it seems to solve our problem. The condition is not satisfied by those who are taken in by Descartes' argument that every event has a cause, since they do not realize that the argument includes a tacit premise they do not accept. It is, however, clearly satisfied for many deductive arguments. The condition is also met by the cases of charting and persistence forecasting, and I have argued that it is also met by the inductive justification of induction, where that argument is taken as one to the degree of reliability of our practices. Since we already accept a method of induction, we may use an inductive argument to assess its reliability, since there could be (indeed there are) optimistic and pessimistic people who would accept both its premises and its rule of inference, yet do not already accept its conclusion. This result accounts for our ambivalence toward Hume's argument against the inductive justification, our feeling that while he has shown that the argument is circular, it nevertheless has some probative value. Circularity is relative to audience, and the inductive justification of induction is circular for an audience of skeptics, yet not among those who already accept that induction is better than guessing.

We may now return to the truth argument, which says we ought to infer first that successful theories are true or approximately true, since this is the best explanation of their success, and then that Inference to the Best Explanation is truth-tropic, since this is the method of inference that guided us to these theories. The objection to this argument that we have been considering is that it is circular, just like the inductive justification of induction. Indeed it might be claimed that the truth argument is just the inductive justification dressed up with a fancy theory of induction. But we have now seen that, while the inductive justification begs against the inductive skeptic, a case can be made for saying that it is not circular in the narrower context of a dispute among those who already accept induction. Similarly, we must concede that the truth argument has no force against those who do not already accept that inferences to the best explanation are truth-tropic. The question before us is whether we may extend the analogy with the inductive justification, and so salvage something from the truth argument by claiming that at least it is not circular for those who already accept a realist version of Inference to the Best Explanation.

One obvious difference between the simple inductive justification of induction and the truth argument is that only the former is enumerative. The inductive justification says that past success provides an argument for future success, but the truth argument says that past success provides an argument for truth, not just for future success. This, however, is just the difference between enumerative induction and Inference to the Best Explanation: what we have are parallel applications of different methods of induction. So the truth argument can be defended against the charge of circularity in the narrower context in the same way as the inductive justification was defended, by showing that the argument can be used to settle disputes. For the inductive justification, the dispute was over degrees of reliability. The truth argument, in a more sophisticated form, could also be used for this dispute. Even successful theories make some false

167

predictions, so the literal truth of the theory cannot quite be the best explanation of its actual track record. More perspicuously, perhaps, the dispute can be seen as one over degree of approximation to the truth. Scientific realists may disagree over how effectively or how quickly scientists get at the truth or over the relative verisimilitude of competing theories, and something like the truth argument could be used to address this sort of dispute.

There is also another dispute that the truth argument might help to settle. In the last chapter, I argued that the distinction between prediction and accommodation is epistemically relevant, because scientists are only imperfect judges of the actual inductive support their theories enjoy. When a theory is constructed to accommodate the data, there is a motive to fudge the theory or the auxiliary statements to make the accommodation. This fudging, if it occurs, reduces the support that the data provide for the theory by the scientist's own standards but, since fudging is not always obvious, she may miss it and so overestimate the credibility of her theory. When a theory makes correct predictions, however, this provides some evidence that the theory was not fudged after all, and so that her assessment was correct. The issue here is not over the reliability of scientists' inductive principles, considered abstractly, but over the reliability of their application. The question is how good scientists are at applying their own principles to concrete cases. This is clearly something over which scientists with the same principles could disagree, and it is a dispute that the truth argument could help to settle, since the predictive success of accepted theories is, among other things, a measure of how well scientists are applying principles that they agree are reliable. So I conclude that, while the truth argument is no argument against the inductive skeptic or the instrumentalist, the circularity objection does not show that realists are not entitled to use it. The argument is circular against non-realists, but not for realists themselves.

A BAD EXPLANATION

To say that the truth explanation does not fall to the circularity objection is not, however, to say that it does not fall. It faces another serious problem, the bad explanation objection. As I have argued in this book, an inference to the best explanation is not simply an inference to what seems the likeliest explanation, but rather the inference that what would be the loveliest explanation is likeliest. And how lovely is truth as an explanation of predictive success? According to the bad explanation objection, not lovely at all. Even if the truth argument is not hopelessly circular, it is still a weak argument, and weak on its own terms. This is so because the argument is supposed to be an inference to the best explanation, but the truth of a theory is not the loveliest available explanation of its predictive success; indeed it may not be an explanation at all. In a way, this is the reverse of the circularity objection. According to that objection, the truth argument is illegitimate precisely because it would be an inference to the best explanation, and so would beg the question, since only a realist could accept it. According to the bad explanation objection, the truth argument is not warranted by Inference to the Best Explanation, so even a realist ought not to accept it. I want to consider two versions of this objection. The first is that there is a lovelier explanation; the second is that, competitors apart, the truth explanation is too ugly to warrant inference.

Van Fraassen has given what he claims is a better explanation of predictive success (1980, pp. 39–40). His explanation is neo-Darwinian: scientific theories tend to have observed consequences that are true because they were selected for precisely that reason. Scientific method provides a selection mechanism that eliminates theories whose observed consequences are false, and this is why the ones that remain tend to have observed consequences that are true. This mechanism makes no appeal to the truth of theories, yet explains the truth of their observed consequences. The realist, however, seems to have a simple response. According to Inference to the Best Explanation, we are to infer the loveliest of *competing* explanations, but the truth explanation and the selection explanation are compatible, so we may infer both. The scientific environment may select for theories with observed consequences that are true and the theories thus selected may be true. Van Fraassen's explanation does not deny that successful theories are true; it just does not affirm this. So it appears that the realist is free to accept van Fraassen's account yet also to make the truth argument.

169

Van Fraassen will surely say that this is too quick. The selection explanation is logically compatible with the truth explanation but, once we infer the selection explanation, we ought to see that it deprives us of any reason we may have thought we had for inferring the truth explanation. The one explanation pre-empts the other. But why should this be? When my computer did not work, I did not infer that the fuse was blown, since I noticed that the computer was unplugged. These two explanations are logically compatible, since they both could be true, but the plug explanation is known on independent grounds to be correct, and it takes away any reason I would have had to infer that the fuse has blown. Once I accept the plug explanation, there is nothing left for the fuse to explain. But I want to argue that this is not the relevant analogy. The realist will accept the selection mechanism, but this does not explain everything that inference to truth would explain.

To see this, notice that a selection mechanism may explain why all the selected objects have a certain feature, without explaining why each of them does (cf. Nozick, 1974, p. 22). If a club only admits members with red hair, that explains why all the members of the club have red hair, but it does not explain why Arthur, who is a member of the club, has red hair. That would perhaps require some genetic account. Similarly, van Fraassen's selection account may explain why all the theories we now accept have been observationally successful, but it does not explain why each of them has been. It does not explain why a particular theory, which was selected for its observational success, has this feature. The truth argument, by contrast, does explain this, by appealing to an intrinsic feature of the theory rather than just to the principle by which it was selected. So the truth argument explains something not accounted for by the selection explanation and this, along with the compatibility of the two accounts, suggests that the selection explanation does not pre-empt the truth explanation.

Yet perhaps even this explanatory difference does not remove the feeling of pre-emption. For the truth explanation was motivated by the thought that the success of science would be miraculous if its theories were not largely true, but the selection explanation seems to remove the miracle. It's no miracle that all the members of the club have red hair, if this was a criterion of selection. But this reaction is misguided. The real miracle is that theories we judge to be well supported go on to make successful

170

predictions. The selection mechanism does not explain this, since it does not explain why our best-supported theories are not refuted in their next application. Constructive empiricism assumes that scientific canons of induction yield theories that will continue to be empirically successful in new applications, but it does not explain why this should happen. The truth explanation, by contrast, does provide some sort of explanation of a theory's continuing predictive success. If our inductive criteria are truth-tropic, then well-supported theories tend to be true, and so they will tend to generate true predictions. This assumes that our criteria are generally truth-tropic, so that is not explained, but the truth argument does explain our continuing observational success and, as we noted in chapter two, an explanation may be sound and provide understanding even though what does the explaining is not itself explained.

An exceptionally assiduous reader may worry about whether what I have just said is in tension with my claims about prediction and accommodation in the last chapter. This paragraph is a detour to address that worry. Roughly speaking, what I have just said is that, while the selection explanation would account for why all the theories we now accept accommodate all the available evidence, only the truth explanation accounts for why they have in fact made so many successful predictions. This, however, seems to assume a stronger epistemic distinction between prediction and accommodation than I found defensible in the last chapter. There I did argue for a principled difference between the two, based on the fudging explanation, but only in assessed and not in actual support, and this depended on our imperfect access to the actual support a theory enjoys. Now, however, we have a much more fundamental difference, since it is only predictions that display the reliability of our inductive methods. But I think I am right both times, because the issues have changed. In the last chapter, we were working entirely within our inferential methods, to see how they might support a distinction between accommodation and prediction. In that context, our imperfect access to actual support was the only defense I could mount for the strong advantage thesis, that a prediction tends to provide more reason to believe a theory than the same datum would have, had it been accommodated. In this chapter, however, we are concerned with assessing those methods themselves, and this is why the epistemic distinction between prediction and accommodation is now more fundamental. Our theories are tested by their accommodations, but our inferential methods are not, since accommodated data are not inferred. (The conflation of these two issues may help to explain why some people think there is a greater epistemic difference between prediction and accommodation in the first, internal case than there really is.)

Back, now, to what the truth explanation explains. The selection explanation and the truth explanation account for different things. The selection explanation accounts for the fact that, at any given time, we only accept (in either the realist's or the constructive empiricist's sense) theories that have not yet been refuted. It assumes nothing about our inductive powers; indeed it is an explanation that Popper might give. The truth explanation, by contrast, accounts for two other facts. First, it explains why a particular theory that was selected is one that has true consequences. Second, it explains why theories that were selected on empirical grounds then went on to more predictive successes. The selection explanation accounts for neither of these facts. I conclude that it does not pre-empt the truth explanation. The truth argument cannot be defeated on the grounds that the selection explanation is a better explanation of observational success and so blocks the inference to truth as the best explanation.

This disposes of the argument that the truth explanation is a bad explanation because the selection explanation is better, but it does not dispose of the bad explanation objection, which can be prosecuted without arguing that there is a better explanation. Even if the truth explanation is nominally the best explanation for continuing observational success, perhaps because it is the only possible explanation for this, it may yet be too feeble to warrant inference. Inference to the Best Explanation, correctly construed, does not warrant inferring the best explanation at all costs; as I have noted before, the best must be good enough to be preferable to no inference at all. Perhaps the likeliest thing is that continuing observational success is just inexplicable, a brute fact. Perhaps it is really no less likely that a false theory should be observationally successful than that a true one should be.

How lovely, then, is the truth explanation? Alas, there is a good reason for saying that it is not lovely at all. The problem is that it is too easy. For any set of observational successes, there are many incompatible theories that would have produced them. This is our old friend, underdetermination. The trouble now is that the truth explanation would apply equally well to any of these theories. In each case, the theory's truth would explain its observational success, and all the explanations are equally lovely. A very complex and *ad hoc* theory provides less lovely explanations than does a simple and unified theory of the same phenomena, but the *truth* of the complex theory is as lovely or ugly an explanation of the *truth* of its predictions as is the explanation that the truth of the simple theory provides. In either case, the explanatory value of the account lies simply in the fact that valid arguments with true premises have only true conclusions. So the truth explanation does not show why we should infer one theory rather than another with the same observed consequences. To appreciate this point, it is important to remember that the proponent of the truth argument holds that his explanation is distinct from the first-order explanatory inferences scientists make. Those inferences, construed as inferences to the best explanation, do distinguish between theories with the same observed consequences, since not every such theory gives an equally lovely explanation of the evidence. This is one of the main strengths of Inference to the Best Explanation as an account of those inferences. But the proponent of the truth argument, as I have construed her, insists that the truth explanation, applied to a particular theory, is distinct from scientific explanations that the theory provides. She is entitled to this, if she wants it. After all, we may suppose that most of those scientific explanations are causal, but the truth explanation is not. The truth of a premise in a valid argument does not cause the conclusion to be true. But the price she pays for this separation is an exceptionally weak explanation, which does not itself show why one theory is more likely than another with the same observed consequences.

If she insists on maintaining the autonomy of the truth explanation, I can only grant the proponent of the truth argument one small consolation. In the face of the circularity objection, she already had to concede that the truth argument only has force for people like herself, who are already realists. These people have already accepted the (approximate) truth of successful theories, on the basis of their construal of the scientists' inferences. For them, the weakness of the truth explanation is not a barrier to

173

inference, since the inference has already been made. So perhaps they may say that the truth argument provides some small additional reason for their position.

THE SCIENTIFIC EVIDENCE

At this point, the better part of valor is for the realist to give up the autonomy claim. The truth explanation can instead be seen as a kind of summary statement of the *scientific* reasons for believing that our best theories are approximately true. If the arguments of this book are along the right lines, these are largely explanatory reasons. But can the realist generate any argument for his position by appeal to the inferences to the best explanation that scientists make? In particular, is there any reason, based on the scientific evidence and the structure of inference, to prefer a realist version of Inference to the Best Explanation over some instrumentalist surrogate, such as van Fraassen's constructive empiricism? This is a big question that I cannot adequately answer here, but I do want to identify some of the realist's resources.

Let us focus on the causal inferences that I have emphasized in this book. Van Fraassen construes these inferences realistically when the inferred causes are observable, but not otherwise. Is there any reason to say instead that all causal inferences are inferences to the actual existence of the causes, observable or not? Nancy Cartwright has argued that we must be realists about causal explanation, because these explanations 'have truth built into them' (1983, p. 91). To accept a causal explanation is to accept that the cause really exists so, insofar as we infer explanations that traffic in unobservable causes, we must also infer the existence of unobservable causes. The idea here is that a causal account only actually explains what it purports to explain if the causal story it tells is true. I agree, but then I am a realist and I find it very implausible to say that any false (and not even approximately true) explanation is an actual explanation, whether the explanation is causal or not. I do not think that you understand why something happens if the story you tell is fiction. Cartwright herself, however, holds that actual explanations need not be true, so long as they are not causal explanations (1983, Essay 8), and she does not succeed in showing why someone like van Fraassen should feel compelled to agree that even causal explanations have truth built into them. Why not say instead that scientists tell causal stories that enable them to make accurate predictions, and that these explanations are actual explanations, so long as they are empirically adequate? Cartwright has not, so far as I can see, shown why the instrumentalist cannot have a non-realist model of causal explanation.

Cartwright's argument also has a peculiar structure. She takes van Fraassen to be challenging the realist to say what it is about the explanatory relation that makes the fact that something would explain a truth 'tend to guarantee' that the explanation is true as well. In the case of causal explanation, she claims, this challenge can be met, because we can only accept the explanation if we accept that the entities and processes it describes actually occur (1983, pp. 4–5). But this is a *non sequitur*. Even if only real causes can explain, this does not show why the fact that a putative cause would explain a truth tends to guarantee that it actually does.

There are, however, some other arguments for saying that someone who, like van Fraassen, construes inferences to observable causes realistically ought to do the same for unobservable causes. I will briefly consider three. The first is the 'same path, no divide' argument. As I have tried to show in earlier chapters, the structure of causal inference is the same, whether the cause is observable or not. Mill's description of the Method of Difference obscures this, since it suggests that we must have independent access to the cause we infer, but Inference to the Best Explanation brings it out, by allowing inference to the existence of causes, as well as to their causal role (cf. chapter seven). So there is a *prima facie* case for saying that all these inferences should be construed in the same way: since we all (in this discussion) grant the truth-tropism of inferences to observable causes, we ought also all to be realists about inferences to unobservable causes, since the inferences have the same form in both cases.

To resist this, the instrumentalist must claim that there is some principled epistemic distinction between inferences to the observable and to the unobservable, even if the paths are the same. On first hearing, this may sound plausible: the existence of the unobservable seems by its nature more speculative than the existence of the observable. This intuition might be strengthened by appeal to underdetermination. Only theories that traffic in unobservables have truth-values underdetermined by all possible evidence. Nevertheless, the claim is misguided. The relevant distinction, if there is one, is between the observed and the unobserved, not between the observable and the unobservable. What counts for our actual epistemic situation is not ideal underdetermination by all possible evidence, but the much greater actual underdetermination by the evidence we now have. But neither the realist nor the instrumentalist is willing to abjure inferences to the truth of the unobserved, since this would make the predictive application of science impossible. To show that scientists are not entitled to infer unobservables, it would at least have to be shown why these inferences are all more precarious than inferences to the observable but unobserved, and no good reason has been given for this.

Inferences to the unobserved are risky but, if they are to the observable, there is at least in principle a way to determine whether they are successful. In cases of inferences to the unobservable, however, it might be claimed that we never can know if we were right, and it is this that makes such inferences intrinsically more speculative than inferences to the observable. Does this line of thought underwrite the epistemic relevance of ideal underdetermination? I think not. It is not clear why an inference we cannot check is therefore illegitimate, or even less likely to be correct than one we can check. Moreover, the realist need not concede that these inferences cannot be checked. To suppose that we cannot go on to test them begs the question. These tests will never prove that a claim about the unobservable is true, but neither will any observation prove that things really are as they appear to be.

Very recently, van Fraassen has argued that the realist advocate of Inference to the Best Explanation must make the unwarranted assumption that a true theory is likely to lie among those that have been considered, or else we should not infer that the best of these is true (1989, p. 143). This is a strange argument for him to use, since a similar argument would apply against the constructive empiricist. He must make the unwarranted assumption that a theory with true observable consequences is likely to lie among those that have been considered, or else we should not infer that the best of these is empirically adequate. Perhaps this assumption is unwarranted, but to abandon it is to endorse extreme inductive skepticism. Since, however, the constructive empiricist is no skeptic, it is not yet clear on what basis he can scout inferences about unobservables. (Of course, a reasonable version of Inference to the Best Explanation does not require that we always infer the best of the available explanations. As I have already noted, the best must be good enough.)

Van Fraassen does not, however, generally argue that inferences to the truth of unobservable claims are always unwarranted, or even less well warranted than inferences to unobserved observables. Instead, he claims that they are unnecessary, on grounds of parsimony (1980, pp. 72ff.). The realist and the constructive empiricist infer exactly the same claims about observables, but the realist also makes additional inferences about unobservables; these additional inferences are unnecessary to the scientist, so he should not make them. The more he infers, the greater the risk he runs of being wrong, so he should only make the inferences he needs. To this, I think the realist ought to give two replies, one easy, one more difficult. The easy one is that he is satisfied if the additional claims he wishes to infer are warranted, whether or not they are strictly required for scientific practice. He may add that, if they can be warranted, such inferences ought to be made, since science is, among other things, the attempt to discover the way the world works, and it is agreed on all sides that the mechanisms of the world are not entirely observable. The more difficult reply is that while the realist does stick his neck out further than the constructive empiricist, simply because he believes more, one of the things he gains is greater warrant than the constructive empiricist for the observable claims they share. If this is so, it is a telling argument for realism; but how could it be?

A scientist who is a constructive empiricist engages in two different types of inference. In the case of observable general claims, he infers the truth of the best explanation, from which he then goes on to deduce predictions. In the case of unobservable theories, however, he only infers the empirical adequacy of the best explanation, and he then also infers the truth of the predictions that follow from them. The realist, by contrast, always employs the first schema, whether the claims are observable or not. Of course the constructive empiricist can cover both his schemata under one description, by saying that he always infers the empirical adequacy of the best explanation, since observable claims are empirically adequate just in case they are true. This, however, leaves the difference, which is simply that, in the first case the lemmas on the road to prediction are believed to be true, while in the second case they are not. Similarly, the constructive empiricist really has two different models of actual explanation. For explanations that appeal only to observables, they must be true to be actual explanations, but for explanations that appeal to unobservables, they must only be empirically adequate. Thus, for observable explanation, all actual explanations must be logically compatible, while for unobservable explanation, actual explanations may be incompatible, though they must have compatible observable consequences. (What should the constructive empiricist say about the many scientific explanations that appeal to both observable and unobservable causes?)

I suggest that the constructive empiricist's bifurcation of both inference and explanation leaves him with less support for his predictive claims from unobservable explanations than the realist has. Call this the 'transfer of support' argument. Consider first the case that the constructive empiricist and the realist share. What reason do they have for believing that the predictions they derive from observable causal explanations that are well supported by the evidence will be true? One reason, perhaps, is that this form of inference has been predictively successful in the past. But there is also another reason. We have found that our method of inferring causes has been a reliable way of discovering actual causes, and predictions deduced from information about actual causes tend to be correct. If the argument of this book is correct, this method prominently includes inferences to the best contrastive explanation and, in the case of inferences to observable causes, we have good reasons to believe that the method is effective. In the cases of causes that are observed but not initially known to be causes, Inference to the Best Explanation inherits the plausibility of the Method of Difference, since in these cases, as we have seen, the two forms of inference have the same structure. In the cases of inferences to observable but unobserved causes, one signal reason we have to trust our method is that we often subsequently observe the causes it led us to infer. If this were not so, we would have substantially less confidence in this extension of the Method of Difference to unobserved (but observable) causes. I infer that the fuse is blown, because none of the lights or electrical appliances in the kitchen is working, and then I go into the basement and see the blown fuse. If we never had this sort of vindication of our causal inferences, we would have much less confidence in them. Fortunately, we enjoy this vindication all the time, and this is something to which both the realist and the constructive empiricist can appeal. But consider now what reason the constructive empiricist has for believing that the predictions he derives from an unobservable causal explanation that is well supported (in his sense) will be true. If he has any reason for this, it can only be that this form of inference to unobservable causes has been predictively successful in the past, a reason the realist has as well. The realist, however, has an additional reason to trust the predictions he generates from the unobservable causes he infers. His reasons for trusting his method in the case of observables also supply him with reasons for trusting his method in the case of unobservables, since it is the same method. In particular, just as his success in sometimes observing the causes he initially inferred supports his confidence in his method when he infers unobserved but observable causes, so it supports his confidence when he infers unobservable causes, because it gives him reason to believe that his method of inference is taking him to actual causes, from whose description the deductions of predictions will tend to be true.

The realist's justified confidence in his predictions comes, in part, from his justified confidence in the existence of the causes his theory postulates, and this confidence comes from his success in subsequently observing some of the causes Inference to the Best Explanation led him to infer. For him, the observed success of Inference to the Best Explanation in locating causes supports the application of that method to unobserved causes and the predictions we generate from their description, whether the causes are unobserved because we were not in the right place at the right time, or because we are constitutionally incapable of observing them. The constructive empiricist cannot transfer this support from the observable to the unobservable case, since he uses a different method of inference in the two cases, and since his method of prediction in the unobservable case does not travel through an inference to the existence of causes. The realist does run a greater risk of believing a falsehood, since he believes so much more, but the benefit of his ambition is that he has better reason than the constructive empiricist to trust his predictions. This argument, if sound, hits the constructive empiricist where it matters most to him, in the grounds for believing claims about the observable but as yet unobserved.

What can the constructive empiricist say in reply? He might try to deny the transfer of support for the realist's method from observable to unobservable inference, but this would require an argument for an epistemic divide between inferences to unobserved observables and inferences to unobservables, an argument that I have suggested has not been provided. Alternatively, he might try to appropriate this transfer of support for his own method of generating predictions by means of unobservable theories. Crudely, he might pretend to be a realist until he makes his prediction, with all the confidence in it that the realist deserves, but then at the last minute cancel his belief in the theory itself. Perhaps this has all the advantages of theft over honest toil, but the accomplished thief does sometimes end up with his booty and out of jail. And this sort of strategy is often perfectly reasonable. I may believe something because it follows from a much larger claim I also believe, but if I can arrange a wager on the consequence alone, I would be silly to bet instead on the larger claim, on the same terms. But the realist need not deny this. It is enough for him to have shown that his confidence in the consequence depends in part on his having good reasons for believing the truth of the larger claim, even if he need not make practical use of that more ambitious inference. For he has shown that he would not have the degree of justified confidence he does have in his predictions, unless he also had good reason to believe his theory, which is what he set out to show.

I want to try out one more argument for realism and against instrumentalism or constructive empiricism, an argument for the initially surprising claim that we sometimes have more reason to believe a theory than we had to believe the evidence that supports it. This claim is not essential to scientific realism but, if it is acceptable, it makes such realism extremely attractive. We may call the argument the 'synergistic' argument. Recall that the 'same path, no divide' argument for realism was in part that the epistemic divide, if there is one, falls between the observed and the unobserved, not between the observed and the unobservable, which is where the instrumentalist needs it. But now I want to question whether all beliefs about unobserved observables are more precarious than beliefs about what is actually observed. One reason for thinking this false is the theory-ladenness of observation. We see with our theories and our dubitable ordinary beliefs, not just with our eyes, so our observational judgments presuppose the truth of claims that go beyond what we have actually observed. In some respects, these observational judgments are like inferences, which it may be possible to analyze as inferences to the best explanation, where the explanations we infer are explanations of our experiences. As Mill remarked, 'in almost every act of our perceiving faculties, observation and inference are intimately blended. What we are said to observe is usually a compound result, of which one-tenth may be observation, and the remaining nine-tenths inference' (1904, IV.I.2).

Observational judgments themselves have an inferential component, so the need for inference to form judgments about the unobserved but observable does not show there to be a principled epistemic divide between the observed and the unobserved. Mill goes on to make a further interesting claim:

> And hence, among other consequences, follows the seeming paradox that a general proposition collected from particulars is often more certainly true than any one of the particular propositions from which, by an act of induction, it was inferred. For each of those particular (or rather singular) propositions involved an inference from the impression on the senses to the fact which caused that impression; and this inference may have been erroneous in any one of the instances, but cannot well have been erroneous in all of them, provided their number was sufficient to eliminate chance. The conclusion, therefore, that is, the general proposition, may deserve more complete reliance than it would be safe to repose in any one of the inductive premises.
>
> (1904, IV.I.2)

This seems to me quite right: we may have more confidence in an inferred generalization than we initially had in the evidence upon which it was inferred. The evidence is never certain, and our justified confidence in it may change. After we infer the generalization, our confidence in each of our data will improve, since each will inherit additional support from the inferred generalization. Moreover, the same point may apply to theories more ambitious than simple generalizations, which appeal to unobservables. When a theory provides a unified explanation of many and diverse observational judgments, and there is no remotely plausible alternative explanation, we may have more confidence in the theory than we had in the conjunction of the evidence from which it was inferred. But if we may have more reason to believe a theory involving unobservables than we initially had for the observations upon which it is based, the claim that we have insufficient reason to infer the truth of a theory would be perverse. The instrumentalist must admit we have sufficient reason to accept our observational judgments, since he wants to say we are entitled to infer the predictions they support, so if we may have even more reason to infer a theory, that is reason enough.

It would be desirable to have detailed historical examples to help make out the claim of the synergistic power of evidence, but I am not the person to provide this. There are, however, simple and interesting cases where I think the claim is plausible. I am thinking of situations were we have reason to believe that the data were fudged, though not fabricated. One example may be Gregor Mendel's pea experiments on the laws of inheritance; another may be Robert Millikan's oil-drop experiments on the discreteness and value of the charge of electrons. In both cases we have good reason to believe there was fudging, in Mendel's case because his data are statistically too good to be true, in Millikan's because we have his laboratory notebooks and so can see how selective he was in using his raw data. One might take the position that what this shows is that Mendel and Millikan actually had no good reason to believe their theories, but I think this is wrong. While each of their observational claims was highly dubitable, taken together they gave more support for their theories than the claims themselves enjoyed before the theories were inferred. In Millikan's case, in particular, his preference for the results on certain days would be unwarranted in the absence of his theory but, given that theory, he may have been reasonable in supposing that the 'bad' data were the result of irrelevant interference with an extremely delicate experiment.

This, then, is a sketch of three promising arguments for realism. Unlike the truth argument, they appeal to the structure of scientific inferences to the best explanation, rather than to an overarching inference of the same form. They also differ from the truth argument in having some force against forms of instrumentalism that grant that we have reason enough to accept both the truth of our observational judgments and the truth of the predictions of well-supported theories, but deny that we have sufficient reason to infer the approximate truth of the theories themselves. The first argument, the 'same structure, no divide' argument, was that we can have the same sort of warrant for claims about unobserved causes as we have for observed causes, since the path of causal inference to the best explanation is the same in both cases and the instrumentalist has not made out a principled epistemic distinction between claims about observables and claims about unobservables. The second, the 'transfer of support' argument, was that only the realist can account for the actual degree of support observable predictions from theories enjoy, since only she can account for the support that the observed successes of inferences to the best explanation in locating observable causes provides for the application of the same form of inference to unobservable causes and so to the predictions we generate from their description. Finally, the 'synergistic' argument was that, because of the inferential structure implicit in our observational judgments and the unifying power of the best explanation, we may have more reason to believe a theory than we initially had for believing the data that support it. I do not expect that the instrumentalists will now all drop dead: instrumentalism, like other forms of skepticism, is a philosophical perennial, and most instrumentalists have already seen variants of the arguments I have just sketched, yet still stand by their positions. These arguments, however, may be enough to show that the failure of the truth argument for realism, at least as an argument against anyone, does not show that all arguments for realism are bound to beg. Moreover, they suggest that, even though the overarching inference to the best explanation upon which the truth explanation relies is too undiscriminating to make a strong case for realism, and even though there are nonrealist versions of Inference to the Best Explanation, the structure of actual causal inferences that Inference to the Best Explanation illuminates shows this account of inference to be a friend to realism after all.

I treat my philosophical intuitions with considerable and perhaps excessive respect. If the intuition refuses to go away, even in the face of an apparently good argument against it, I either look for further arguments that will make it go away, or I find a way to defend it. Endorsing a philosophy I cannot believe does not interest me. This book is a case in point. Some years ago, I wrote a dissertation attacking Inference to the Best Explanation, but my belief that the account is fundamentally right would not go away, so now I have written a book defending it. Or witness my heroic defense in the last chapter of the view that predictions are better than accommodations, in spite of the apparently strong arguments on the other side. I could not get rid of the belief that predictions are better, so I found an argument. But what about the truth argument? In the first two sections of this chapter, we found that it is almost entirely without probative force. Yet I am left with the intuition that underlies it. If I were a scientist, and my theory explained extensive and varied evidence, and there was no alternative explanation that was nearly as lovely, I would find it irresistible to infer that my theory was approximately true. It would seem miraculous that the theory should have these explanatory successes, yet not have something importantly true about it. As Darwin said about his theory, 'It can hardly be supposed that a false theory would explain, in so satisfactory a manner as does the theory of natural selection, the several large classes of facts above specified' (Darwin, 1859, p. 476). But this intuition does not depend on the truth argument. It is not that the truth of the theory is the best explanation of its explanatory or predictive success; it is simply that the theory provides the best explanations of the phenomena that the evidence describes. We find inference compelling in such a case because we are creatures that judge likeliness on explanatory grounds. This is why Inference to the Best Explanation gives a good account of our actual inferential practices. This does not show that all arguments for scientific realism must beg the question, but it does suggest that, in the end, the best evidence for scientific realism is the scientific evidence, and the structure of the methods scientists use to draw their inferences from it.

CONCLUSION

This inquiry into Inference to the Best Explanation has been an attempt at both articulation and defense. The project of articulation was pursued in two ways: directly, by giving more content to the model, and indirectly, by confronting it with a battery of challenges. I began by distinguishing the problems of justification and description and by making a case for the importance and autonomy of the descriptive project. The history of epistemology has been driven by skeptical arguments, and this has resulted in what is perhaps an excessive focus on justification and a relative neglect of the apparently more mundane project of principled description. I also urged a certain conception of the descriptive project, one that focuses on the deductive underdetermination of our inferences by the evidence available to us and attempts to discover the black box mechanisms that we actually employ to generate determinate inferences. The model of Inference to the Best Explanation is to be seen as a putative description of one of these mechanisms, perhaps the central one.

Turning to the model itself, I attempted a direct articulation in three main ways. First of all, I urged the importance of the distinction between potential and actual explanation. Inference to the Best Explanation must be Inference to the Best Potential Explanation, instances of which are inferences that the best potential explanation is an actual explanation. Without a conception of potential explanation, the account would not be epistemically effective and could not explain how explanatory considerations could be a guide to the truth; with it, we can see more clearly just what the model can and cannot provide by way of solutions to problems of justification. Second, I made the distinction between the member of the pool of candidate explanations most war-

185

ranted by the evidence – the likeliest explanation – and the member which would, if true, provide the most understanding – the loveliest explanation. The model tends to triviality if we understand 'best' as likeliest, since the sources of our judgments of likeliness are precisely what the model is supposed to illuminate. Inference to the Loveliest Explanation, by contrast, captures the central idea that the explanatory virtues are our guides to inference, so I urged that we construe the model in this ambitious and interesting form. Finally, I attempted to give some specific content to the notion of loveliest explanation, primarily through my analysis of contrastive explanation, but also by appeal to the notions of causal mechanism, precision, and unification. The project of articulation was also pursued by presenting the model with a series of challenges. There turn out to be six in all. The first is to give an informative account of what makes one explanation better than another. The second is to show that inferences to the best explanation are more than inferences to the likeliest cause. The third and fourth are to show that the model marks a substantial improvement on the hypothetico-deductive model of confirmation and on Mill's methods. The fifth is to answer Hungerford's objection that, since explanatory beauty is in the eye of the beholder, loveliness cannot provide a suitably objective guide to inference. Finally, the sixth challenge is to answer Voltaire's objection, that the model is too good to be true since, even if loveliness is objective, we have no reason to believe that we inhabit the loveliest of all possible worlds, no reason to believe that the explanation that would provide the most understanding is also the explanation that is likeliest to be true.

My defense of Inference to the Best Explanation as a partial description of the mechanisms that govern our inductive practices can be seen primarily as an attempt to begin to meet these six challenges. The challenge from the obscurity of explanatory loveliness was addressed primarily by my account of contrastive explanation, as I have already mentioned. The structural similarity between that account and Mill's Method of Difference also enabled me to argue that Inference to the Best Explanation can be more than Inference to the Likeliest Cause since, by looking for explanations that would exclude appropriate foils, we are led by explanatory considerations to what are in fact likeliest causes. This idea was developed by a detailed examination of the structure of Ignaz Semmelweis's research into the cause of childbed fever. This discussion also helped to meet Hempel's challenge, by showing how Inference to the Best Explanation can illuminate the context of discovery, an issue that the hypothetico-deductive model explicitly ignores, and by showing how the explanationist model can register relevant evidence that the deductive model misses, where there are no plausible deductive connections between theory and data. I then went on to argue that Inference to the Best Explanation also avoids some of the over-permissiveness of the deductive model, and considered this in detail for the case of the raven paradox.

Since my argument for the superiority of the explanationist model over the deductive model depended so heavily on the similarity between inferences to contrastive explanations and applications of the Method of Difference, this made Mill's challenge particularly pressing. I suggested two advantages over Mill. Unlike Mill's methods, Inference to the Best Explanation applies to the wide range of cases where what is inferred is the existence and not merely the causal status of a cause; cases that, while obviously analogous to applications of the Method of Difference, are not strictly warranted by that method. I also pointed out that the Method of Difference does not itself help to solve the problem of multiple differences, which is to show how we decide which of many prior differences between fact and foil are causally relevant, a problem that Mill's own assumptions make inevitable. In a very preliminary way, I then went on to suggest that the considerations we use to solve this problem are ones that contribute to explanatory loveliness and that we finesse much of the problem through the short-list mechanism that inferences to the best explanation must employ.

This leaves Hungerford's and Voltaire's objections. My reply to the challenge that loveliness is too variable to be our guide to inference was that, on the one hand, there is substantial inferential variability and that, on the other, my contrastive analysis of explanation shows how what counts as a good explanation can be genuinely interest relative without thereby being subjective in a sense that would make explanatory considerations unsuitable guides to inference. In response to Voltaire's objection, that the success of Inference to the Best Explanation would be miraculous, I observed that the success of Mill's methods is, at one level of skepticism, no miracle, and, in light of my account of contrastive explanation, this shows that the success of inferences to the best explanation would be no miracle either. At a Humean level of skepticism, of course, the success of any inductive policy is miraculous, but from this point of view Voltaire's objection does not succeed in showing that the explanationist model of inference is less plausible than any other. I also suggested that the match between explanatory and inferential considerations need not be fortuitous, since it is plausible to hold that our explanatory standards are malleable and could have been molded to serve as guides to successful inferences.

The case for Inference to the Best Explanation must of course also rest on its power to give a natural description of many and various sorts of inference we actually make. I hope that the examples scattered through the book help to show this. One of the main attractions of the model is that it accounts in a natural and unified way both for the inferences to unobservable entities and processes that characterize much scientific research and for many of the mundane inferences about middle-sized dry goods that we make every day. It is also to its credit that the model gives a natural account of its own discovery, that the model may itself be the best available explanation of our inductive behavior since, as we have seen, that inference must itself be inductive and moreover an inference to a largely unobservable mechanism. Furthermore, the model offers a satisfying explanatory unification of our inductive and explanatory practices, and one that casts light on the point of explanation and the reason this activity occupies so large a part of our cognitive lives.

In the last two chapters of this book, I turned to the question of whether the explanationist model helps to solve some of the problems of inductive justification. The results were mixed. It can be used to justify particular aspects of our inferential practices, such as our preference for predictions over accommodations; and the structure of scientific inference that it reveals can be used to provide some arguments for adopting a realist rather than an instrumentalist stance towards scientific theories. We found, however, that the attempt to argue that predictively successful theories are likely to be approximately true, on the grounds that the truth of a theory is the best explanation of its predictive success, is not promising, since it begs against the opponents of scientific realism and does not provide a good enough explanation even to supply those who are already realists with much additional support for their position. Moreover, I do not

188

now see how Inference to the Best Explanation helps us to solve Hume's problem of induction. The model does offer a description of our inductive behavior that is importantly different from the simple extrapolation model that Hume assumed, and so perhaps undermines the theory of habit formation that Hume himself adopted as part of his 'skeptical solution', but his skeptical problem seems discouragingly insensitive to the particular description we favor.

I am acutely aware that this book has only scratched the surface of our inductive practices or even of one particular model of those practices, both because of the hardness of the surface and the softness of my nails. It has included many arguments, undoubtedly of unequal value. I suppose that I am most satisfied with my discussion of contrastive explanation and the way this can be applied to articulate and defend the explanationist model of inference, least satisfied with my discussion of the role of unification and other explanatory considerations in guiding inference. Certainly much more needs to be said about what makes one explanation better than another and about which aspects of our inductive behavior the explanationist model can cover and which it cannot. Also, while I have defended the model, I also believe that it has generally enjoyed rather more support than it really deserves, since it has remained for so long little more than an attractive slogan, in part because of the general neglect to distinguish clearly between the plausible but relatively banal view that we make inductions by means of inferences to the likeliest explanation and the exciting but by no means obviously correct view that we employ inferences to the loveliest explanation. Nevertheless, I hope that I have said enough to convince some readers that the exciting view merits further development. I also take some comfort in the otherwise discouraging fact that an account of our inductive practices does not have to be very good to be the best we now have.

BIBLIOGRAPHY

Austin, J. L. (1962) *Sense and Sensibilia*, Oxford: Oxford University Press.

Ayer, A. J. (1956) *The Problem of Knowledge*, Harmondsworth: Penguin.

Black, M. (1970) 'Induction and Experience', in L. Foster and J. W. Swanson (eds) *Experience and Theory*, pp. 135–60, London: Duckworth.

Braithwaite, R. B. (1953) *Scientific Explanation*, Cambridge: Cambridge University Press.

Brody, B. (1970) 'Confirmation and Explanation', in B. Brody (ed.) *Readings in the Philosophy of Science*, pp. 410–26, Englewood Cliffs: Prentice Hall.

Bromberger, S. (1966) 'Why-Questions', in R. G. Colodny (ed.) *Mind and Cosmos*, pp. 86–108, Pittsburgh: University of Pittsburgh Press.

Campbell, D. (1974) 'Evolutionary Epistemology', in P. A. Schilpp (ed.) *The Philosophy of Karl Popper*, pp. 413–63, LaSalle: Open Court.

Cartwright, N. (1983) *How the Laws of Physics Lie*, Oxford: Oxford University Press.

Chomsky, N. (1965) *Aspects of the Theory of Syntax*, Cambridge, Mass.: MIT Press.

Chomsky, N. (1986) *Knowledge of Language*, New York: Praeger.

Churchland, P. M. and Hooker, C. A. (eds) (1985) *Images of Science*, Chicago: University of Chicago Press.

Cummins, R. (1989) *Meaning and Mental Representation*, Cambridge, Mass.: MIT Press.

Darwin, C. (1859) *On the Origin of Species*, New York: Collier (1962).

Descartes, R. (1641) *Meditations on First Philosophy*, trans. by Donald Cress, Indianapolis: Hackett (1979).

Dreske, F. (1972) 'Contrastive Statements', *The Philosophical Review* 82: 411–37.

Fine, A. (1984) 'The Natural Ontological Attitude', in J. Leplin (ed.) *Scientific Realism*, pp. 83–107, Berkeley: University of California Press.

Friedman, M. (1974) 'Explanation and Scientific Understanding', *The Journal of Philosophy* 71: 1–19.

Garfinkel, A. (1981) *Forms of Explanation*, New Haven: Yale University Press.

Gettier, E. (1963) 'Is Justified True Belief Knowledge?', *Analysis* 23: 121–3.

190

Gjertsen, D. (1989) *Science and Philosophy: Past and Present*, Harmondsworth: Penguin.
Glymour, C. (1980) *Theory and Evidence*, Princeton: Princeton University Press.
Goodman, N. (1983) *Fact, Fiction and Forecast*, 4th edn, Indianapolis: Bobbs-Merrill.
Hanson, N. R. (1972) *Patterns of Discovery*, Cambridge: Cambridge University Press.
Harman, G. (1965) 'The Inference to the Best Explanation', *The Philosophical Review* 74: 88–95.
Harman, G. (1973) *Thought*, Princeton: Princeton University Press.
Harman, G. (1986) *Change in View*, Cambridge, Mass.: MIT Press.
Hempel, C. (1965) *Aspects of Scientific Explanation*, New York: Free Press.
Hempel, C. (1966) *The Philosophy of Natural Science*, Englewood Cliffs: Prentice-Hall.
Horwich, P. (1982) *Probability and Evidence*, Cambridge: Cambridge University Press.
Hume, D. (1777) *Enquiries Concerning Human Understanding and Concerning the Principles of Morals*, L. A. Selby-Bigg and P. H. Nidditch (eds) Oxford: Oxford University Press (1975).
Jevons, W. S. (1877) *Elementary Lessons in Logic*, 6th edn, London: Macmillan.
Kuhn, T. (1970) *The Structure of Scientific Revolutions*, 2nd edn, Chicago: University of Chicago Press.
Kuhn, T. (1977) *The Essential Tension*, Chicago: University of Chicago Press.
Laudan, L. (1984) 'A Confutation of Convergent Realism', in J. Lepin (ed.) *Scientific Realism*, pp. 218–49, Berkeley: University of California Press.
Lepin, J. (ed.) (1984) *Scientific Realism*, Berkeley: University of California Press.
Lewis, D. (1986) 'Causal Explanation', in *Philosophical Papers*, Vol. II, pp. 214–40, New York: Oxford University Press.
Lipton, P. (1987) 'A Real Contrast', *Analysis* 47: 207–8.
Lipton, P. (1990) 'Prediction and Prejudice', *International Studies in the Philosophy of Science* 4(1): 51–65.
Lipton, P. (1991) 'Contrastive Explanation', in D. Knowles, (ed.) *Explanation and its Limits*, Cambridge: Cambridge University Press.
Mackie, J. L. (1974) *The Cement of the Universe*, Oxford: Oxford University Press.
Maher, P. (1988) 'Prediction, Accommodation, and the Logic of Discovery', in A. Fine and J. Lepin (eds) *Philosophy of Science Association 1988*, vol. 1, pp. 273–85, East Lansing: Philosophy of Science Association.
Mill, J. S. (1904) *A System of Logic*, 8th edn, London: Longmans, Green and Co.
Nozick, R. (1974) *Anarchy, State, and Utopia*, New York: Basic Books.
Nozick, R. (1983) 'Simplicity as Fall-Out', in L. Cauman et al. (eds) *How Many Questions?*, pp. 105–19, Indianapolis: Hackett.

Pierce, C. S. (1931) *Collected Papers*, (eds) C. Hartshorn and P. Weiss, Cambridge, Mass.: Harvard University Press.

Popper, K. (1959) *The Logic of Scientific Discovery*, London: Hutchinson.

Popper, K. (1962) *Conjectures and Refutations*, London: Routledge & Kegan Paul.

Popper, K. (1972) *Objective Knowledge*, Oxford: Oxford University Press.

Putnam, H. (1975) *Mathematics, Matter, and Method*, Cambridge: Cambridge University Press.

Putnam, H. (1978) *Meaning and the Moral Sciences*, London: Hutchinson.

Quine, W. v. O. (1951) 'Two Dogmas of Empiricism', *The Philosophical Review* 60: 20–43.

Quine, W. v. O. (1969) 'Natural Kinds', in *Ontological Relativity* pp. 114–38, New York: Columbia University Press.

Quine, W. v. O. and Ullian, J. (1978) *The Web of Belief*, 2nd edn, New York: Random House.

Ruben, D. (1987) 'Explaining Contrastive Facts', *Analysis* 47 (1): 35–7.

Schlesinger, G. (1987) 'Accommodation and Prediction', *Australasian Journal of Philosophy* 65: 33–42.

Skyrms, B. (1986) *Choice and Chance*, 3rd edn, Belmont: Wadsworth.

Sober, E. (1986) 'Explanatory Presupposition', *Australasian Journal of Philosophy* 64 (2): 143–9.

Stein, E. and Lipton, P. (1989) 'Evolutionary Epistemology and the Anomaly of Guided Variation', *Biology and Philosophy* 4: 33–56.

Temple, D. (1988) 'The Contrast Theory of Why-Questions', *Philosophy of Science* 55 (1): 141–51.

Thagard, P. (1978) 'The Best Explanation: Criteria for Theory Choice', *The Journal of Philosophy* 75: 76–92.

van Fraassen, B. C. (1980) *The Scientific Image*, Oxford: Oxford University Press.

van Fraassen, B. C. (1989) *Laws and Symmetry*, Oxford: Oxford University Press.

Williams, B. (1978) *Descartes*, Harmondsworth: Penguin.

Woodward, J. (1984) 'A Theory of Singular Causal Explanation', *Erkenntnis* 21: 231–62.

Worrall, J. (1984) 'An Unreal Image', *The British Journal for the Philosophy of Science* 35 (1): 65–80.

INDEX

accommodation and prediction 3, 13, 71, 73, 123–4, 133–57, 168, 171–2
actual and potential explanation 3, 59–61, 74, 160, 185
Austin, J.L. 150

Black, M. 164
Brody, B. 58
Bromberger, S. 30

Cartwright, N. 58, 63, 74, 174–5
Chomsky, N. 6–8, 15–16
context of discovery 69, 88–90, 117, 119–22, 138, 145–6, 152–3, 156
Cummins, R. 102

Darwin, Charles 184
Darwinian evolution 120–2, 130–1
deductive-nomological model of explanation 2, 29–33, 38–41, 45, 50–5, 58, 60, 118
Descartes, René 9–11, 162, 166
difference condition on contrastive explanation 43–55, 76–7
Digby, Sir Kenelm 117

familiarity model of explanation 27–9, 32, 54
Fine, A. 4, 160
Freud, S. 40

Friedman, M. 24, 27, 30, 53, 119

Garfinkel, A. 35–7, 43
Gettier cases 60
Gjertsen, D. 117
Glymour, C. 19
Goodman, N. 17, 19, 99–103, 105

Hanson, N. R. 58
Harman, G. 58, 60, 66, 122
Hempel, C. 16, 18, 26–7, 29, 31, 35–7, 40, 60, 65, 67, 79–80, 88–9, 91, 95, 100–1, 105
higher-level inferences to the best explanation 70–3, 84, 86, 90, 125, 133–84
Horwich, P. 71–3, 135, 138, 150
Howson, C. 144
Hume, David ix, 10–14, 16–17, 20, 24–5, 126, 161–2, 167
Hungerford's objection 73, 122–5, 156
Hussey, T. 116
hypothetico-deductive model of confirmation 3, 18–20, 30–1, 58, 63, 67–71, 75, 77, 88–97, 99–100, 112

instantial model of confirmation 16–8, 69–70
interest relativity of explanation 2, 28–9, 35, 46, 51–2, 54, 75–7, 88, 124–5

Kuhn, T. 7–8, 15, 71–2, 123, 129, 131, 151

Laudan, L. 4, 160
Lewis, D. 32, 42–3, 53
likeliest and loveliest explanation 3, 59, 61–4, 66, 71, 73–7, 116–19, 122–32, 169, 186

Maher, P. 134
Mendel G. 182
Mendeleyev, D. 134
Mill, J. S. 20, 43, 96, 109–10, 115, 181
Mill's methods of induction 20–1, 43, 77–8, 96, 104, 109–17, 125, 128, 175, 178
Millikan, Robert 182

negative evidence 80–6, 91–5
new riddle of induction 17, 19, 100–1
Newtonian mechanics 52, 62
non-causal explanation 33–4
Nozick, R. 109, 137, 170

partial explanation 40
preadaptation 120–2
Pierce, C. S. 58
Popper, K. 94, 126, 144–5, 160
Putnam, H. 159

Quine, W. v. O. 102, 105, 122, 146

raven paradox 17–19, 31, 69, 78, 100–12

realism 59, 72–3, 126–7, 158–84
reason model of explanation 26–7, 32, 56
Ruben, D. 36

Schapiro, M. 149
Schlesinger, G. 135
self-evidencing explanations 26–7, 30, 32, 56–7
Semmelweis, I. 68, 79–98, 103–5, 115, 117, 139
Skyrms, B. 12, 161
Sober, E. 50
Stein, E. 121

Temple, D. 36, 39
Thagard, P. 58
theory-ladenness of observation 180–1

Ullian, J. 122
underdetermination 6–8, 12, 15, 22, 31, 34, 146, 173, 175–6, 185
unification 71, 119, 121

van Fraassen B. 30, 35, 37, 126–8, 160–1, 169–72, 174–84
Voltaire' objection 74, 122–3, 125–32

Williams, B. 9
Williamson, T. 51
why regress 24, 32
Worral, J. 51

Zemach, E. 45